BEYOND THE WHITE COAT

✳ greenhill

(✳) greenhill
https://greenhillpublishing.com.au/

Mulcahy, Liam (author)
Beyond The White Coat: The Faces Behind Medical Eponyms
ISBN: 978-1-923523-73-9 (paperback)
ISBN: 978-1-923589-36-0 (hardcover)
MEDICAL HISTORY
NON-FICTION

Cover Image : AI Generated (Midjourney)
Typesetting Calluna Regular 11/16
Cover and book design by Green Hill Publishing

Beyond the White Coat

The Faces Behind Medical Eponyms

LIAM MULCAHY

"The good physician treats the disease;
the great physician treats the patient who has the disease."

— Sir William Osler

Contents

A Note to the Reader

THIS BOOK EXPLORES the lives of individuals whose names have become immortalised through medical eponyms. While their contributions to medicine are undeniably significant, their lives and work were deeply shaped by the historical and cultural contexts of their time. Some content may touch on sensitive topics or describe practices, beliefs, and attitudes that are now considered offensive or outdated. These accounts are presented not to endorse such perspectives, but to offer a more nuanced understanding of the eras in which these individuals lived and worked.

This is not a comprehensive history of medicine, nor an exhaustive biography of the figures discussed. Rather, it highlights compelling moments in their lives, discoveries, and legacies. In some instances, where the historical record is silent or limited, I've used informed imagination to bring key moments to life—crafting scenes or dialogue that remain true to the spirit of the evidence while allowing room for emotional and narrative insight.

This book offers a glimpse into the medical practices and breakthroughs of the past—and how they continue to shape modern medicine.

Finally, please note that this is not intended as medical advice. Descriptions of treatments and diagnoses reflect historical approaches and should not be applied to contemporary medical decision-making. Readers should approach this as a historical narrative, not medical reference.

Thank you for joining me on this journey into the lives of these remarkable figures in medical history.

Enjoy!

What is a Medical Eponym?

"Are you telling me that you built a time machine... out of a DeLorean?"

IF YOU GREW UP in the 1980s—or have a soft spot for nostalgic films—this line might instantly transport you back in time to *Back to the Future*. The iconic sci-fi film didn't just make us dream of time travel; it also turned Michael J. Fox into a global star. With his boyish charm, quick wit, and magnetic presence, Fox became a symbol of youthful energy and Hollywood success.

But in 1991, at the age of 29 and the height of his career, Fox noticed a twitch in his pinky finger. At first, it seemed harmless—perhaps just stress or fatigue from a gruelling filming schedule. But the tremor persisted. As other subtle symptoms emerged, he sought medical advice. The diagnosis? *Parkinson's disease.*

It was devastating. *Parkinson's*—a progressive neurological disorder—was something Fox had associated with much older people. How could this happen to someone so young, so vibrant? For years, he kept the diagnosis private, grappling with what it meant for his future. But in 1998, he made the courageous decision to go public, becoming one of the most prominent advocates for *Parkinson's* research. His foundation, established shortly after, has since become the world's largest nonprofit funder of *Parkinson's* research.

Stories like Fox's are how many of us first encounter certain medical conditions—through the lives of public figures. Muhammad Ali's battle with *Parkinson's* became as well-known as his boxing legacy. More recently, singer Lewis Capaldi has opened up about his experiences with *Tourette's syndrome*, shining a light on a condition many people didn't understand.

But these names go beyond celebrity headlines. Perhaps you've watched a loved one slowly fade from *Alzheimer's disease* or experienced the joys and challenges of raising a child with *Down syndrome*. These names are examples of medical eponyms.

So, what is an eponym?

The word "eponym" comes from the Greek *epi* (upon) and *onyma* (name). At its core, an eponym is a name given to something—a discovery, invention, place, or idea—in honour of a person associated with it. And though we may not think about them often, eponyms are everywhere.

Consider the diesel engine—named after Rudolf Diesel, the German engineer who patented it in 1892. It revolutionised transport and industry. Or the word *boycott*, derived from Charles Boycott, an English land agent ostracised by Irish tenants in the late 19th century during a political protest. His name came to define the act of organised refusal.

Even fashion has its eponyms. The cardigan was named after James Brudenell, the 7th Earl of Cardigan, who popularised the buttoned garment during the Crimean War. And no, the Caesar salad wasn't named after Julius Caesar—it was created by Caesar Cardini, an Italian-American restaurateur in the 1920s.

And then there's America itself—named after Amerigo Vespucci, the Italian explorer whose accounts of the New World helped mapmakers define the continents. His name became one of the most recognisable eponyms of all.

As these examples show, eponyms go far beyond science or medicine. They mark moments in history, innovations, and the people behind them. But in medicine, they carry a particular poignancy.

If you've ever been to a hospital or GP, you've probably come across a medical eponym without even realising it.

A medical eponym is more than a label; it's a link to the past. Behind each one is a person—often forgotten—whose work was shaped by the world around them. These aren't just names. They're stories—complex, messy, and beautifully human.

One doctor found himself in the middle of an assassination attempt on a king. Another saved lives by hacking apart the only lifeboat on a sinking ship. Some endured unthinkable hardship, like the physician imprisoned in a Nazi POW camp who held onto his humanity amidst horror. Others were struck by tragedy—one shot by a patient.

Not all stories are dramatic. Some are defined by innovation in extraordinary circumstances—like the surgeon who, amidst the chaos of World War I, developed pioneering neurosurgical techniques while treating wounded soldiers in the trenches. His work would transform battlefield medicine—and neurosurgery itself.

These glimpses reveal that medical eponyms aren't just scientific shorthand. They're windows into the lives of individuals shaped by war, culture, politics, and personal conviction.

This book was born from a desire to uncover those stories—the remarkable, often forgotten lives behind names we hear so often in medicine. Names like *Alzheimer*, *Down*, and *Parkinson* have become part of our everyday language, yet the people behind them remain largely unknown. Their discoveries shaped how we understand and treat disease—but their personal journeys are just as compelling.

This book is for everyone—whether you're a medical professional, a history enthusiast, or simply curious about the people behind the names. It's a journey through history, told through human stories as layered and intricate as the conditions they helped define.

So as you turn the pages, I invite you to look beyond the white coat—to see the humanity, struggle, brilliance, and complexity behind the eponyms. The next time you hear a medical condition named after someone, I hope you'll think not only of the disease—but of the person whose name endures.

CHAPTER 1

The Rebellious Movement Disorder

IN 1794, DURING the reign of George III, Britain was a nation on edge. The American colonies had broken away, leaving behind a shaken empire, and the fiery ideals of the French Revolution were spreading both fear and inspiration across Europe. Amidst this climate of upheaval, a covert scheme known as the *Pop-Gun Plot* emerged, threatening the very life of the monarch.

The plot was audacious. Conceived by members of a radical reformist group, they believed that the only way to instigate real change was to eliminate King George III. The conspirators plotted to assassinate the King using a poisoned dart fired from an airgun— a device that looked more like a child's toy than a weapon of revolution, hence the name '*Pop-Gun*'.

One dark evening, a man was dragged from his home, his protests drowned beneath the heavy boots of soldiers. Neighbours stood in stunned silence as he was hauled away to the Tower of London. His supposed crime? Alleged involvement in the *Pop-Gun Plot*. For now, his name was whispered in scandalous rumours, but history would remember him for far nobler reasons. As he walked through the Tower's shadowed halls, the chill of the stone walls and the low echoes of footsteps seemed to taunt him with the fate awaiting traitors to the crown. Yet beneath the fear, he clung to his principles.

A few days later, he stood before the Privy Council—the King's most trusted advisors. The air in the chamber was thick with tension, every eye fixed on the prisoner. At the head of the room sat William Pitt the Younger, the sharp-eyed, and calculating Prime Minister, who regarded the accused in silence.

"Pray, sir," Pitt began, his voice clipped and deliberate, "how came you to be invited onto this committee?"

The prisoner stood tall, his voice calm. "Because, Mr Pitt, they did me the honour to believe I was firm in the cause of Parliamentary reform. I had just published a tract for the benefit of the wives and children of those imprisoned on charges of high treason."

A ripple of murmurs passed through the chamber. His writings had become a beacon for those demanding justice—a cry for universal suffrage, annual parliaments, and equity in a society entrenched in privilege. To some, he was a symbol of progress. To others, a dangerous agitator.

Pitt leaned forward slightly; his tone more pointed. "And what of your involvement in this so-called *Pop-Gun Plot*?"

The man met his gaze without flinching. "I refuse to testify until I am assured I will not be compelled to incriminate myself."

Pitt's faint smile faded, replaced by something that looked like reluctant respect. The interrogation dragged on, but the prisoner stood his ground. He offered nothing more—no names, no admissions. His belief in reform and his quiet defiance shone brighter than ever. It stunned many to see such composure in a physician known for his kindness, now caught in the web of political conspiracy.

That man, who stood unshaken before the Privy Council, was James Parkinson—the pioneering doctor who would one day give his name to *Parkinson's disease*.

In the mid-18th century, Hoxton Square in Middlesex, England, was a lively and close-knit neighbourhood. Rows of terraced red-brick houses lined the cobbled streets, their decorative windows and doors giving the square a distinctive charm. Life there hummed with quiet rhythm—residents exchanged greetings, ran errands, and found comfort in the gentle predictability of the community. At its heart stood a modest schoolhouse where local children learnt their letters and sums. Nearby, a workhouse provided refuge for the destitute, offering food, shelter, and labour to those with nowhere else to turn. It was a place of hardship and resilience, a small world where people faced challenges together.

It was here, at No. 1 Hoxton Square, that James Parkinson was born on 11th April 1755. His father, John Parkinson, was a respected apothecary and surgeon whose skill and care earned him the trust of the community. From an early age, James was drawn to his father's work, watching with fascination as he mixed tinctures (alcohol-based medicinal extracts), diagnosed ailments, and tended to patients who arrived at their door. For the young boy, home was more than a house—it was a place where science and humanity met, where healing began not only with medicine but with understanding.

Parkinson's childhood was steeped in curiosity and compassion. He spent hours by his father's side, absorbing the art of medicine. The scent of crushed herbs, the careful compounding of remedies, and the quiet conversations with the sick became etched into his memory. These early experiences left a profound impression, shaping his sense of responsibility towards others.

Intent on following in his father's footsteps, Parkinson began a rigorous seven-year apprenticeship under his guidance. It was demanding work, but it transformed him into a skilled surgeon and pharmacist. Despite his success and stability, his intellectual appetite stretched beyond Hoxton. At 20, he enrolled in a six-month programme at the *London Hospital Medical College* in Mile End, one of the era's most progressive institutions. There, he encountered new medical theories and cutting-edge techniques that enriched the practical foundation built at home. By 1784, at the age of 29, James Parkinson qualified as a surgeon—well-equipped to serve not only with expertise, but empathy.

Parkinson was more than a physician; he was a man attuned to the struggles of everyday people. He was disturbed by how many of the illnesses he encountered were preventable—yet the knowledge needed to prevent them remained out of reach for the poor. That injustice lit a fire in him. He began to write simple, practical health guides designed not for scholars, but for ordinary people. These pamphlets offered guidance on everything from child-rearing to hygiene, and the dangers of overindulgence to basic home remedies. They aimed to empower families with the tools to take control of their health, bridging the chasm between medicine and those who needed it most.

His desire to help extended beyond paper. In 1777, Parkinson was awarded a silver medal by the *Royal Humane Society* for saving the life of a man who had attempted to hang himself. Using a resuscitative technique that combined physical intervention with electric shocks—a bold and experimental approach at the time—he revived the man from the brink of death. This wasn't just about innovation; it reflected his conviction that every life mattered.

His compassion didn't stop at physical ailments. For over three decades, Parkinson served as the visiting doctor for *Holly House*, one of Hoxton's three private asylums. At a time when mental illness was poorly understood and often stigmatised, Parkinson treated his patients with empathy and patience. Where others saw lost causes, he saw individuals deserving of dignity. His commitment to humane care, long before it was fashionable or even expected, set him apart.

One of his most impactful works was *The Town and Country Friend and Physician*, a book written in accessible language that addressed common medical concerns. It covered everything from caring for infants to treating rabies, epilepsy, and intestinal worms. It urged readers to inoculate against smallpox—a groundbreaking and life-saving measure at the time. Every page carried the same underlying message: knowledge was power, and health should not be a privilege reserved for the wealthy. The book became a lifeline for families who lacked access to private physicians.

As the Industrial Revolution took hold, Hoxton Square began to change. The once serene and picturesque neighbourhood under-went tremendous change. Factories rose where open spaces once stood, and the rhythms of daily life quickened to match the pace of progress. While innovation brought economic growth, it also deep-ened the divide between rich and poor. Parkinson couldn't ignore the inequality spreading around him. As he walked the streets of a rapidly changing London, he grew uneasy.

Industrial progress had lifted some, but it left many more behind. Parkinson refused to remain silent. He felt a duty to speak for those without a voice—and he found a way to do so without risking his livelihood. Despite his many accomplishments and his reputation as a respected physician, Parkinson harboured a secret identity, one that allowed him to voice his most radical ideas.

In the shadows, James Parkinson became *'Old Hubert.'*

The fervour of the French Revolution ignited a spark in James Parkinson that would shape his political convictions. He was extremely critical of the British government, particularly under Prime Minister William Pitt the Younger, whom he saw as indif-ferent to the struggles of ordinary citizens. In Parkinson's eyes, those in power were insulated by privilege and paralysed by self-interest.

By night, away from his medical practice, he sat by candlelight, pouring his frustrations and hopes onto the page. His pamphlets called for sweeping reforms—universal suffrage, improved education for the poor, and humane treatment of prisoners. These writings resonated with the disillusioned and disenfranchised, but they also put him at considerable personal risk. To protect himself and his family, he adopted the pseudonym *'Old Hubert.'*

As *Old Hubert*, Parkinson wrote with sharp wit and unflinching criticism. His commentaries were bold, biting, and laced with satire—making them both popular and dangerous. He called out government abuses, condemned corruption, and championed the idea that knowledge and political power should not be confined to the elite. His words found their way into reformist circles, where they inspired action—and into government offices, where they stirred suspicion.

Parkinson became active in the *London Corresponding Society* and the *Society for Constitutional Information*, two leading groups pushing for electoral reform. Both were under close surveillance, and their members were frequently harassed or arrested. *Old Hubert's* essays became a vital voice within these movements, amplifying the call for justice and democratic representation.

Then came the *Pop-Gun Plot.*

The alleged conspiracy to assassinate King George III with a poisoned airgun was sensational, if not absurd. Several members of the *London Corresponding Society* were arrested, and by association, Parkinson found himself implicated. The government seized the opportunity to strike at reformists under the guise of protecting the crown.

The atmosphere was thick with suspicion as Parkinson was summoned before the Privy Council. Accusations hung in the air like smoke. Prime Minister Pitt himself led the questioning; his piercing gaze fixed on the physician-turned-activist.

"Pray, Dr Parkinson, how came you to be invited onto this committee?" he asked, his voice steely.

Parkinson, composed but inwardly anxious, replied, "Because I believe they did me the honour to believe me firm in the cause of Parliamentary reform—and because I had just published a little tract for the benefit of the wives and children of the persons imprisoned on charges of high treason."

The room stilled. His calm defiance was unmistakable. But Parkinson was acutely aware of the danger he faced. What if they refused to believe him? What if his family suffered for his politics?

Still, he stood firm. He refused to incriminate himself, stating clearly that he would not testify without a guarantee of legal protection. Ultimately, no charges were brought against him, though several of his associates endured months in prison before being acquitted.

The *Pop-Gun Plot*, historians later concluded, was likely exaggerated or fabricated—a convenient tool for suppressing reformist dissent. Though Parkinson emerged physically unscathed and grateful to see daylight once more, the ordeal left a mark. It reminded him of the lengths to which those in power would go to silence critics. But it did not silence him. As *Old Hubert*, he continued to write. His pen remained sharp. His cause remained just.

As London marched into the 19th century, so too did Parkinson's focus shift—from political activism back to medicine. The crowded streets of Hoxton Square echoed the city's transformation, and amidst the growing noise and industry, Parkinson found a renewed sense of purpose in his clinical work. The passion that had once fuelled his radical writings as *"Old Hubert"* now powered his commitment to improving public health. His medical interests were broad and forward-thinking, including some of the earliest clinical accounts of appendicitis (inflammation of the appendix) and peritonitis (inflammation of the abdominal lining). These illnesses were often misunderstood and deadly at the time.

One afternoon, Parkinson strolled through the familiar streets of Hoxton Square. The square buzzed with its usual activity as he greeted locals with warm nods and exchanged brief conversations. Children played near the schoolhouse, their laughter weaving between the cries of vendors hawking their goods. Nearby, shop-keepers beckoned passersby into their stores, while neighbours gathered in doorways to share gossip and greetings. In the midst of this vibrant, lived-in scene, Parkinson's attention was drawn to an elderly man making his way through the crowd with evident difficulty.

The man's hunched posture and shuffling gait immediately caught Parkinson's trained eye. His steps were slow and uncer-tain. Even as he stood still, his limbs trembled. His face wore a look of quiet frustration—tinged with resignation—the expression of someone betrayed by his own body.

Parkinson's curiosity was piqued. As he continued his walk, the streets he knew so well began to take on a strange new weight. He saw another man, and then another, displaying similar symp-toms: the same stooped posture, the same hesitant stride, the same tremor. Their movements were like a tragic choreography—bodies struggling against invisible restraints, each step an effort.

Approaching these men with the gentle authority of an experi-enced physician, Parkinson introduced himself and began asking questions. He listened attentively as they described their experiences, and slowly a pattern emerged. None of the men knew one another, yet they told similar stories—of trembling limbs, reduced strength, and a tendency to lean forward when walking. They described being able to think clearly but feeling trapped inside faltering bodies. Each man had once lived with strength and independence. Now, they battled with frustration, embarrassment, and loss of control.

Back in his clinic, Parkinson encountered more patients with this strange condition. His modest but busy practice in Hoxton Square became a quiet observatory for a medical mystery. At first, he speculated that these tremors might originate from damage to the cervical spine—a theory that fit with prevailing medical thought. But the consistency and specificity of the symptoms led him to question that conclusion. No existing diagnosis matched what he was seeing. Parkinson began to suspect he was observing something entirely new.

Driven by his deep-seated belief in observation and detail, he began keeping records. He documented not only the physical symptoms but also how they progressed over time—the subtle stiffening of muscles, the involuntary movements, the change in pace and posture. His notes were precise, clinical, yet full of empathy. These were not just cases; they were people, and Parkinson was determined to understand what was happening to them.

In 1817, at the age of 62, he published his findings in a slim but groundbreaking work titled *An Essay on the Shaking Palsy*. In it, he described the disease with remarkable clarity:

"Involuntary tremulous motion, with lessened muscular power, in parts not in action and even when supported; with a propensity to bend the trunk forwards, and to pass from a walking to a running pace: the senses and intellects being uninjured."

It was the first clinical description of what would later become known as *Parkinson's disease.*

The pamphlet received quiet respect within the English medical community but did not garner wide recognition. Parkinson, who had never sought personal fame, returned to his daily practice. But one can imagine his frustration—knowing he had captured something important, yet watching it pass unnoticed by most of his peers. Still, he pressed on, caring for his patients with the same attentiveness, even as he continued to refine his understanding of the condition.

A few years later, Parkinson suffered a stroke that left him mute and partially paralysed. He died on 21st December 1824, at the age of 69. Though he did not live to see the true impact of his work, his legacy was far from lost.

Decades later, in Paris, the celebrated neurologist Jean-Martin Charcot encountered patients whose symptoms perfectly matched Parkinson's descriptions. Struck by the clarity and detail of the essay, Charcot praised the work as a model of clinical observation. He proposed renaming the condition *"Parkinson's disease"* in honour of the man who first captured its essence.

Charcot's endorsement ensured that James Parkinson's quiet brilliance would not be forgotten. Once a modest pamphlet by a socially minded physician, the essay became a foundational text in the study of neurological disorders—a lasting tribute to one doctor's ability to see clearly, listen closely, and give voice to the silent struggles of his patients.

What is Parkinson's disease?

Parkinson's disease is a condition that gradually worsens over time, affecting the nervous system and, most notably, movement. It is characterised by symptoms like tremors, slowness of movement, muscle stiffness, and problems with balance. These occur because specific brain cells that produce dopamine, a chemical that helps control smooth and coordinated movements, are lost. Without enough dopamine, tasks like walking, writing, or standing steadily can become increasingly difficult. Although there is currently no cure, a combination of medication, physical therapies, and surgical interventions—such as deep brain stimulation—can help manage symptoms and improve quality of life.

CHAPTER 2

Catch Your Breath

IT WAS A mild evening in late spring, May 2016. *Deupree House* in Cincinnati, Ohio, hummed with the familiar rhythms of daily life. This senior living community was a sanctuary for its residents, offering just the right mix of independence and support in a warm, inviting atmosphere. Outside, the carefully tended gardens were vibrant with colour, while inside, the elegant common areas provided a tranquil backdrop for both lively conversations and quiet reflection.

As the clock neared 7:00p.m., the dining hall began to fill with its usual crowd. The room, with its expansive windows framing the lush gardens, came alive with the chatter and laughter of friends catching up over their evening meal. Among them was 87-year-old Patty Ris, whose warm smile and quick wit had made her a favourite in the community. For Patty, these dinners were more than just routine—they were a cherished part of her life, a chance to connect and share stories.

She slid into her usual spot that evening, chatting easily with the gentleman beside her. The atmosphere was as comforting as the food, reflecting the life she'd built there. Around her, staff moved deftly, balancing attentiveness with discretion as they ensured each table had what it needed.

Dinner was served, and Patty took a bite of her hamburger, savouring the first taste. But something was wrong.

Her expression twisted as she clutched at her throat, her chair scraping harshly against the floor as she pushed back. She stood abruptly, panic overtaking her features. She couldn't speak. She couldn't breathe.

For a moment, the dining room froze in collective shock. Forks clattered onto plates. Conversations stopped mid-sentence. Patty's face flushed a deep crimson, her wide, panicked eyes darting helplessly. Friends seated nearby stared in disbelief, immobilised by the sudden turn of events.

The 96-year-old gentleman beside her rose calmly, his movements deliberate and composed. Stepping behind her, he clasped his hands below her ribcage and delivered a sharp, upward thrust. The soft thud of his hands against her abdomen reverberated in the silence. Patty's body jolted—but the obstruction held.

Another thrust. Then another.

Finally, a piece of meat with a tiny bone attached shot from her mouth, bouncing off her plate with a startling clink. The room seemed to exhale all at once. Applause broke out, mingled with gasps of relief and astonishment.

Patty gasped for air, coughing and spluttering as the colour returned to her face. The gentleman steadied her in her chair and offered her a glass of water, remaining by her side until he was sure she was all right. Then, with quiet composure, he returned to his seat as the room filled with murmured awe.

Perry Gaines, the maître d', stood rooted to the spot, barely able to process what he'd witnessed. "When I saw who it was, I knew it was historic," he later said. "At his age, that's a very physical type of activity. To see him do it is a fascinating thing. The whole dining room—you could hear a needle drop."

The applause faded, replaced by low voices buzzing with disbelief. They had just witnessed something extraordinary: an act of heroism that would be spoken of for years.

The man who had saved Patty's life was the very one who had invented the life-saving manoeuvre over four decades earlier.

Dr Henry Heimlich.

Henry Heimlich was born to Mary and Philip Heimlich in February 1920, in Wilmington, Delaware. His grandparents, Hungarian and Russian Jewish immigrants, had arrived in America with little more than hope and resilience, planting the seeds of ambition that would run through the family's story. The Heimlichs later moved to New Rochelle, New York, where Henry's curious, inventive nature began to take root.

Philip Heimlich worked as a social worker, dedicating his life to supporting prisoners and troubled youth. Young Henry often accompanied his father on rounds, wide-eyed, absorbing stories of hardship and redemption. These early experiences left a lasting impression, offering him a glimpse into the transformative power of compassion and ingenuity. His mother, Mary, had a quiet but unyielding strength. She'd raised her younger siblings alone after their mother's death—a resilience that would come to shape her son's moral compass.

Even as a child, Henry's imagination seemed boundless. At five, he fashioned a sword from a broken umbrella and paraded through the neighbourhood with grand tales of heroism. He'd spend hours by the stream near his home, fishing with a makeshift rod and a paperclip for a hook. He rarely caught anything, but he didn't mind. His sister, Cecelia, often said that Henry never minded coming up empty—he was too busy dreaming.

His fascination with problem-solving and helping others guided him towards medicine. After graduating from *New Rochelle High School* in 1937, he enrolled at *Cornell University*, where he would become the Big Red Marching Band's drum major. In 1941, he earned his BA, and by 1943, at just 23 years old, he had received his M.D. from *Weill Cornell Medical College*.

Not long after, the world was at war— and Heimlich was drawn into its orbit, he enlisted in the U.S. Navy, eager to serve. His adventurous spirit led him to volunteer for a mission in China as part of the *Sino-American Special Technical Cooperative Organisation (SACO)*, a small but crucial unit tasked with gathering weather intelligence and aiding guerrilla resistance against Japanese forces.

Stationed at Camp Four, near the edge of the Gobi Desert in northern China, Heimlich served as chief medical officer, treating both American and Chinese personnel. Before long, he became the unofficial doctor for a nearby village, where his skills were put to constant test. One of the most urgent problems was *trachoma*—a bacterial eye infection that, left untreated, could lead to blindness. The existing treatments were ineffective, but Heimlich was not one to yield. He devised a new remedy by blending an antibiotic with shaving cream—a strange but inspired concoction. It worked. His improvised solution saved countless villagers from losing their sight. For Heimlich, it was a revelation: the simplest ideas, born of necessity, could yield extraordinary results.

But not all problems had such clean resolutions.

One night in 1945, as the war edged towards its end, a Chinese soldier was brought to the camp with a gunshot wound to the chest. Heimlich and his team worked frantically in the dim operating room, the air thick with tension, antiseptic, and the soldier's ragged breathing. They tried everything—every technique Heimlich knew. But the wound was too grave. The young soldier died on the table, his face pale and still beneath the flickering light.

Heimlich stood there, his gloves slick with blood, the weight of failure pressing on his shoulders. It was his first loss—a brutal reminder of the limits of knowledge and skill in the face of war's raw violence. The moment etched itself into his mind.

The next day, seeking some semblance of peace, he rode a horse toward a nearby town. As he neared the outskirts, he saw an oxcart making its slow way along the dusty road. The soldier's body lay in the back, wrapped and still. The sight caught him. The oxcart—humble, steady, ancient—seemed to embody the harsh finality of life and death in wartime.

Heimlich sat motionless, watching the cart vanish into the distance, its wheels creaking softly with each turn.

If only I'd known more, he thought. *If only I'd had the right tools.*

That moment became a quiet vow. Alone with the dust and silence, Heimlich promised himself he would do better. He would find the tools, the knowledge—whatever it took—to keep that kind of loss from happening again. The soldier's death was no longer just a memory. It became a spark.

After the war, Henry Heimlich's life pivoted sharply towards innovation. He became one of the first surgeons in the United States to be certified in thoracic surgery—a field that fascinated him. He was particularly drawn to the mysteries of lung physiology and the deadly mechanics of pneumothorax, a condition in which air or blood compresses the lung following a puncture of the chest wall.

His passion for solving medical problems soon found its way into papers and proposals that began to attract attention. In 1955, he introduced a bold surgical concept: using a section of a patient's stomach to reconstruct a damaged oesophagus—the muscular tube that carries food from throat to stomach. It was a radical idea, and in the United States, it barely caused a ripple. But across the Atlantic, one man was listening.

Dr Dan Gavriliu, a Romanian surgeon, had been performing a similar operation for years. Intrigued and encouraged by Gavriliu's work, Heimlich travelled to Bucharest to observe, collaborate, and refine his approach. He returned to New York armed with new insights—and successfully introduced the procedure to American surgical practice.

Yet, despite his rising reputation, Heimlich remained haunted by the memory of the Chinese soldier who had died on his table during the war. The image stayed with him, and sometimes, in the quieter hours, he would take long walks through New York's streets, letting the noise of the city muffle the weight of that memory.

On one such walk, he wandered into a small Japanese shop brimming with colourful trinkets. He wasn't looking for anything in particular, but a simple noisemaker caught his eye—bright, light, and whimsical. It contained a flutter valve that produced a soft, rhythmic sound when air passed through it. The design was elementary, even childlike. But to Heimlich, it held a kind of mechanical poetry.

He turned it over in his hands. The simplicity of the mechanism lit a spark.

Back in his lab, the memory of the soldier sharpened his focus. What if such a valve could be adapted into a medical device? What if it could allow air and blood to escape a chest wound—but prevent anything from flowing back in? In the chaos of a battlefield, or a rural village, such a device could mean the difference between life and death.

Over the following weeks, Heimlich got to work. He modified the toy's valve, attaching it to a drainage tube and ensuring it could be sterilised. The final result was elegantly simple: a one-way chest drain valve, light enough to be carried in a field kit, requiring no bulky machinery or suction devices. It allowed air or fluid to escape the pleural cavity, giving a collapsed lung the chance to re-expand and function again.

When the *Heimlich Valve* debuted in the early 1960s, its practicality and effectiveness quickly became apparent. During the Vietnam War, medics from both sides carried it into combat. Soldiers, civilians, and guerrilla fighters alike benefited from its life-saving utility. On battlefields far removed from hospital wards, the device proved itself, over and over again.

Decades later, in 1993, Heimlich returned to Vietnam. There, he met a gathering of thoracic surgeons who had come together for one reason: to thank him. They shared stories of patients saved, of emergencies averted, of moments when the valve had turned the tide. Heimlich listened, quietly moved. For him, it was more than recognition—it was vindication. A full-circle moment.

One evening in 1972, Henry Heimlich came across a series of news stories that troubled him deeply—reports of people choking to death in restaurants, surrounded by helpless bystanders. The thought of diners dying in public, their final moments unfolding amid clinking cutlery and startled silence disturbed him. He believed these deaths were preventable. He had already saved lives with his chest valve; now he felt a moral imperative to act again. If there was a solution, he would find it.

Heimlich returned to his lab with renewed urgency—and a 38-pound beagle.

The dog was partially anaesthetised, and Heimlich reassured his team that it was sedated, not harmed. To simulate choking, he inserted a tube into its throat. His assistant fetched a piece of meat from the hospital kitchen, which Heimlich lodged in the tube to create a grimly realistic obstruction.

At first, he tried pressing on the dog's chest, hoping to force the object out. Nothing happened. The situation grew tense as the dog remained on the edge of suffocation. The room fell silent, save for the sound of its laboured, failing breaths. Frustrated, Heimlich removed the tube and stepped back, mind racing.

Then, in a flash of insight, he considered the diaphragm—the dome-shaped muscle beneath the lungs. It struck him that a quick, forceful push beneath the ribcage might force air upwards through the lungs, creating enough pressure to dislodge the blockage.

Acting on instinct, he balled his fists and delivered an upward thrust into the dog's abdomen.

The piece of meat shot out.

His team watched in stunned silence. He repeated the experiment on three more dogs. Each time, the result was the same—the airway cleared, the animal revived.

The *Heimlich manoeuvre* had been born.

Thrilled by the discovery, Heimlich knew he had found a way to save lives. But now came the harder task: convincing others to use it.

He wrote an article titled *"Pop Goes the Café Coronary,"* published in the June 1974 issue of *Emergency Medicine.* In it, he described the manoeuvre and invited readers to try it in real emergencies—and to write back with their experiences.

Just days after the article appeared, it was put to the test. A retired restaurant owner in Washington State read about the technique and used it to save his choking neighbour. The story made the *Seattle Times* under the headline *"News Article Helps Prevent a Choking Death."* Similar rescues soon followed, each one reinforcing the effectiveness—and simplicity—of the method.

Still, the medical establishment remained sceptical. The *American Red Cross* and the *American Heart Association* initially recommended back blows as the first response, relegating the *Heimlich manoeuvre* to a secondary option. Heimlich was furious.

"They're death blows," he said bluntly, arguing that back slaps could worsen the obstruction. He was convinced, so sure, in his belief that abdominal thrusts were the safer, more effective method.

When medical institutions hesitated, Heimlich took his case to the public. He sent copies of his article to newspapers, appeared on radio programmes, and booked himself on television shows, including *The Tonight Show* with Johnny Carson. On air, he demonstrated the manoeuvre with celebrities like Angie Dickinson, showcasing its ease and effectiveness. His charisma, —coupled with the mounting number of dramatic rescues, —transformed him into a household name.

By 1985, his persistence paid off. The *American Heart Association* formally endorsed the *Heimlich manoeuvre* as the primary treatment for choking. A year later, it was recognised globally and taught in first aid courses around the world.

Testimonials flooded in. A five-year-old in Massachusetts saved a playmate after seeing the manoeuvre on television. Teachers, parents, restaurant staff—ordinary people, empowered with the knowledge, were saving lives.

One of the most powerful endorsements came from Surgeon General C. Everett Koop, who declared that the *Heimlich manoeuvre* should be the *only* method used to treat choking victims. Heimlich's invention had achieved its aim: it placed life-saving power in the hands of the public.

He capitalised on the momentum by founding the *Heimlich Institute*, dedicated to spreading awareness of the technique. The Institute produced training videos, pamphlets, and public education campaigns. His name became inseparable from the manoeuvre itself—a rare case in medicine where the eponym was not just scientific, but iconic.

The numbers were staggering. The *Heimlich Institute* estimated that over 50,000 lives had been saved in the United States alone. A *New York Times* article in 2009 placed the global figure closer to 100,000. Even the *American Medical Association*, once cautious, acknowledged the manoeuvre's immense value.

There were some concerns. Critics warned of potential injuries from abdominal thrusts—broken ribs, internal trauma. In 2006, both the *American Red Cross* and the *American Heart Association* revised their guidelines, reintroducing back blows alongside abdominal thrusts. The term *"Heimlich manoeuvre"* was quietly retired from official recommendations.

Still, despite shifts in policy, the technique remained widely recognised and taught. And compared to Heimlich's later ventures, it would remain his least controversial invention.

In the early 1980s, Dr Henry Heimlich turned his attention to what would become the most contentious chapter of his career. At the centre of it was a radical idea: *malariotherapy*—deliberately infecting patients with malaria to treat diseases such as cancer, Lyme disease, and even HIV.

The concept was not entirely new. In the early 20th century, before antibiotics, malariotherapy had been used to treat advanced syphilis, with some success. The high fevers induced by malaria were thought to kill off the syphilitic bacteria. But the method had long since been abandoned, deemed dangerous, unpredictable, and largely obsolete.

Heimlich, however, believed the principle still had merit. He argued that the intense immune response triggered by malaria could, in theory, help the body fight other diseases. It was the fever that interested him—the body pushed to its biological limits, turned into its own weapon.

The medical establishment, however, did not share his enthusiasm. The *U.S. Food and Drug Administration* and the *Centres for Disease*

Control and Prevention condemned malariotherapy as unscientific and dangerous. Human rights advocates called it "atrocious." Ethicists raised alarm over the deliberate infection of vulnerable patients. Despite these warnings, Heimlich pressed on.

The *Heimlich Institute* began conducting trials in Ethiopia—without informing the country's Ministry of Health. Heimlich claimed that initial tests on seven subjects showed promising results. But when pressed, he refused to release detailed data or medical documentation. Meanwhile, existing studies suggested that malaria could *worsen* some of the very conditions he hoped to treat—especially HIV. The secrecy surrounding the trials, coupled with Heimlich's unwavering confidence in the face of mounting evidence, deepened the controversy, leading many to question whether he was blinded by ambition rather than science.

Around the same time, he proposed another unorthodox idea: applying the *Heimlich manoeuvre* to drowning victims.

He suggested that abdominal thrusts could force water out of the lungs, potentially reviving victims who might otherwise drown. But medical experts strongly disagreed. Drowning, they argued, was caused primarily by oxygen deprivation—not water accumulation in the lungs. The manoeuvre, they warned, could cause serious internal injuries without addressing the root problem. Like malariotherapy, this idea found little traction in the scientific community—and further damaged Heimlich's credibility.

As the criticisms mounted, Heimlich's public image began to unravel. The man once hailed as a life-saving genius was now viewed by many of his peers as reckless, even dangerous. The *Heimlich manoeuvre* remained celebrated—but it was no longer enough to shield him from scrutiny.

Then, the accusations became personal.

Letters and emails began arriving at medical organisations, universities, and media outlets. They accused Heimlich of professional misconduct, of taking undue credit for the manoeuvre, and of engaging in illegal human experimentation. The complaints came from names like "Dr Bob Smith," "David Ionescu," and "Holly Martins." They were persistent, articulate, and damning.

The *University of California* launched an inquiry into collaborations between its researchers and the *Heimlich Institute*, uncovering violations of federal guidelines. The *Cincinnati Enquirer* published a scathing front-page article in which a rival physician described Heimlich as "a liar and a thief." Even the *American Red Cross*—once an early supporter of the *Heimlich manoeuvre*—began to quietly reassess its position.

The barrage of accusations left the Heimlich family reeling.

"It's an incredibly painful and difficult thing for someone to go through in the twilight of his life," said his eldest son, Phil Heimlich. Determined to uncover the source of what he called a "hate campaign," Heimlich hired a lawyer and a private investigator.

What they discovered would shatter the family from within.

The investigation was long and frustrating, often ending in dead ends. The complaints had come from multiple email addresses, different phone numbers, names with academic credentials. All were fakes. Whoever was behind the letters had taken great care to remain anonymous—but they had made one mistake.

Despite the aliases, all the complaints could be traced back to the same internet service provider. "Dr Bob Smith," "David Ionescu," and "Holly Martins" were not a team of whistleblowers. They were one person. A single voice.

That discovery narrowed the field. But the most shocking revelation was still to come.

As the lawyer and investigator continued their search, they began to notice something else—a pattern in the tone and phrasing of the letters. There was something oddly familiar in the voice. Something Heimlich himself couldn't quite put his finger on.

In a final effort, the investigator widened the digital net, scanning obscure online postings linked to the same phone numbers. One of them led to an old, classified advertisement in Portland, Oregon— someone selling a 27-inch television and VCR. The ad had been signed with a single name.

Pete.

The realisation landed with brutal clarity.

The anonymous figure behind the years-long smear campaign— the relentless source of complaints, accusations, and reputational sabotage—was not a professional rival. Not a disillusioned colleague. Not a stranger at all.

It was Peter Heimlich, Henry's own son.

Peter's crusade against his father had begun, at least in his mind, as a search for justice. He believed that Henry Heimlich had ignored significant medical issues within their family—grievances that festered over the years. Fuelled by a growing sense of betrayal, Peter started digging into his father's past. At first, it was a personal investigation. But it quickly became something more.

He combed through academic journals, old newspaper archives, and scientific papers, searching for inconsistencies, contradictions, anything that might support what he increasingly saw as a fraudulent legacy. The estrangement between father and son only deepened. Peter eventually shut down his own business to focus entirely on the mission. What began as a vendetta transformed into something, at least in his own eyes, like an ethical crusade.

Peter filed formal complaints with organisations such as the *Institute of Medicine* and the *National Academy of Sciences*. When they failed to act, he accused them of shielding his father and turned instead to the media. Casting himself as a modern-day David challenging the Goliath of Henry Heimlich's reputation, he attracted the attention of a few journalists. His campaign even influenced the *American Red Cross* to revise its First Aid guidelines, reducing the prominence of the *Heimlich manoeuvre.*

But not everyone saw him as a crusader.

Phil Heimlich, Peter's older brother, remained loyal to their father and publicly condemned Peter's actions as "inappropriate and abusive." The rift between the brothers mirrored the deeper fracture within the family—a painful collision of personal history, clashing values, and the complex legacies we inherit and resist.

As Peter broadened his campaign, he launched a website dedicated to exposing what he considered medical fraud. He expanded his efforts beyond his father, publishing stories and commentary on other alleged ethical violations in the medical field. But none of it had the impact of his sustained, targeted efforts against Henry Heimlich.

For the Heimlich family, the damage was devastatingly personal— and irreparable.

Whatever one's view of Henry Heimlich's later work, the impact of his most famous invention was undeniable. The *Heimlich manoeuvre* had saved thousands—perhaps hundreds of thousands—of lives. It became a cultural touchstone, immortalised in television shows, cartoons, and school curricula. It was simple. It was accessible. It worked.

In May 2016, months before his death, Heimlich stepped from the pages of medical history and back into the present—saving Patty Ris's life with the technique that bore his name. The moment brought him back into the public eye, if only briefly, as news outlets retold the story of the doctor who had once again performed the manoeuvre himself.

He and Patty remained close afterwards, bonded by that strange, life-affirming twist of fate.

Just a few months later, on 17th December 2016, Dr Henry Heimlich died of a heart attack at the age of 96.

The timing felt poetic. He had ended his life as he had lived it— helping someone breathe again.

What is the Heimlich manoeuvre?

The *Heimlich manoeuvre* was a first aid technique used to help someone choking due to an object blocking their airway, such as food or another obstruction. This potentially life-saving procedure was performed to restore breathing and prevent unconsciousness by expelling the object from the airway.

The technique involved standing behind the choking person and wrapping your arms around their waist. A fist would be made with one hand and placed just above the belly button but below the ribcage. The other hand would then grasp the fist, and a series of quick, upward thrusts into the abdomen would be delivered. These thrusts increased pressure in the chest, which could force the object out of the airway. While it was widely taught as a first aid procedure, its application has since been adapted, and guidelines for treating choking may vary today.

CHAPTER 3

Tics and Turmoil

THE LATE AFTERNOON sun slanted through the tall windows of Georges Gilles de la Tourette's consulting room, casting golden light across rows of books, their spines alternating between brilliance and shadow. He sat hunched over his desk, his pen scratching paper in steady, rhythmic strokes. The quiet was dense, almost oppressive, broken only by the soft rustle of turning pages.

But the stillness no longer brought comfort. Not anymore.

An ache lodged deep in his chest, dull and persistent. His son, Jean, stolen by meningitis. His mentor, Jean-Martin Charcot, gone within months of that loss. Grief clung to him like damp wool. Jean's laughter, Charcot's steady voice—their absence was not just painful, but disorienting. He'd survived their loss through sheer force of will. But now, as he tried to write, the words blurred, lost their meaning. His thoughts wandered, fragmented, dissolving on the page.

Then the door slammed open.

A woman stumbled in. Her hair was wild, her face pale and tight with fury. Her eyes locked on him—feral, unhinged—and something in their depths made his heart seize.

Tourette's pen halted mid-sentence. A dark inkblot bloomed across the page.

"Madame?" He rose slowly, voice calm but laced with unease.

"You!" she spat, her words sharp and splintered. "You did this to me—you and your damned hypnotists!"

Her hand darted beneath her coat. A pistol flashed in the lamp-light, the barrel catching a sliver of sun—a gleaming thread of dread.

Tourette froze, hand raised, palm outward.

"Please," he said, steady, though fear twisted in his gut.

She didn't hesitate.

The shot cracked like a whip in the enclosed room. Pain exploded in his neck—hot, searing—as he stumbled backwards and crashed to the floor. His hands flew to the wound, warm blood spilling through his fingers. He gasped, choking on air that no longer came freely.

The world wavered. Light dimmed.

His son, Jean. His mentor, Charcot. Their faces shimmered behind his eyes, sharp against the haze. He had withstood so much, held on through heartbreak and loss.

But now—lying on the floor of his own consulting room, blood pooling beneath him—he wasn't sure he had anything left to hold on *with*.

The room tilted. His vision closed in at the edges.

So, is this how it ends?

Nine years earlier, a young Georges Gilles de la Tourette moved swiftly through the crowded streets of Paris, heart pounding with anticipation. He had arrived in the City of Light in 1881 to continue his medical training at *Laennec Hospital*. Now, in 1884, aged 26, he was finally on his way to witness what he had long dreamt of: Jean-Martin Charcot's legendary Tuesday lecture at the *Salpêtrière*.

This was no ordinary class. It was an event—a spectacle that attracted not only doctors and scientists, but the intellectual elite and curious high society of Paris.

As he walked, Tourette allowed his thoughts to drift. Born in the small town of Saint-Gervais-les-Trois-Clochers in 1857, he had been the eldest of four, always restless, always inquisitive. At 16, he began studying medicine in *Poitiers*, pouring himself into it with fierce focus. By 1881, he'd made it to Paris—the beating heart of modern medicine. He had trained under the likes of Damaschino and Fournier, but no one captured his imagination like Charcot.

And now, at last, he would see him in person.

The *Salpêtrière* loomed ahead—a vast, imposing complex that straddled grandeur and melancholy. Once a grim asylum for the poor and mentally ill, it had, under Charcot's direction, transformed into a hub of medical innovation. Tourette joined the stream of visitors making their way across the gravel courtyard, their footsteps and quiet conversation adding to the sense of anticipation.

Inside the amphitheatre, he paused, momentarily overcome. The room was packed to the rafters, alive with whispered speculation and eager curiosity. Writers, artists, aristocrats, journalists—all crammed shoulder to shoulder, Parisian society at its most curious and theatrical. Some were drawn by genuine intellectual interest. Others came, quite plainly, for the spectacle.

Tourette found a seat near the front. He didn't want to miss a moment.

As he waited, he absorbed every detail—the murmur of the crowd, the thick anticipation in the air, the mixture of perfume and candle smoke. The *Salpêtrière* was more than a hospital. It was a stage where the human mind, in all its mystery, would be laid bare.

The hour struck. A hush fell. Charcot stepped into the room.

He had presence—not just the authority of rank, but a gravitational pull. His gaze swept the amphitheatre with sharp, clinical awareness, like a man taking in not an audience, but a field of data.

"Today," he began, his voice clear and precise, "we will explore *grande hystérie* and the stages of hypnotism: lethargy, cataplexy, and somnambulism."

What followed was as astonishing as Tourette had imagined.

A female patient was brought forward. Under Charcot's direction, she progressed through each hypnotic stage, her body and mind yielding visibly to suggestion.

First came *lethargy*—her muscles slackened, limbs heavy, as if she had fallen into an unnatural sleep. Charcot described it clinically, yet his tone hinted at something reverent. Then came *cataplexy*—a sudden, complete collapse of muscle tone triggered on command. Her body crumpled to the floor, as if a marionette's strings had been cut. Finally, she entered *somnambulism*—a waking dream state. She rose slowly, eyes glassy, arms outstretched in search of something unseen.

Around him, the audience whispered and gasped. But Tourette didn't hear them. He was transfixed—not just by the patient, but by Charcot. The way he orchestrated the session, balancing scientific precision with dramatic tension. He wasn't just demonstrating symptoms—he was *mapping* the unknown.

This wasn't theatre for Tourette. It was a revelation. Charcot wasn't simply observing the mind—he was daring to *understand* it. To bring order to what others dismissed as madness, possession, hysteria.

Tourette felt something shift inside him.

As the lesson ended, the room erupted in applause. Charcot gave a slight nod, already turning back to his notes. Tourette rose from his seat. He made his way to the stage, weaving through the dispersing crowd. His heart thudded.

"Professor Charcot," he said, voice steady, though his nerves surged beneath it. "My name is Georges Gilles de la Tourette. I've recently begun my internship here at the *Salpêtrière*. It's an honour to meet you."

Charcot turned. His gaze, close up, was intense—as though it might strip a man to his constituent parts.

"Ah, Gilles de la Tourette," he said, slowly. "I've heard of you. What did you make of today's lesson?"

Tourette hesitated—for a moment—then replied, "It was extraordinary. Your work on hysteria and hypnotism is revolutionary. To see it in person... it's something I'll never forget. I hope to learn from you—and contribute, if I can."

Charcot studied him, expression unreadable. Then, a brief nod, his stern face softening slightly. A flicker of approval.

"We shall see what you are capable of," he said. "There is much to be done. I expect excellence."

"Of course, Professor," Tourette replied, bowing slightly.

Charcot returned to his notes without another word.

As Tourette left the amphitheatre, the noise of the city seemed to recede behind him. A clarity had settled. This was more than a career. It was a calling. And he would give it everything he had.

�⚕

In the weeks following his meeting with Charcot, Georges Gilles de la Tourette threw himself into his new assignment with a characteristic blend of intensity and curiosity. His task was as peculiar as it was fascinating: to investigate reports of unusual reflex disorders from across the globe—conditions known as the Jumping Frenchmen of Maine, *myriachit*, and *latah*.

The story of the Jumping Frenchmen began in the dense, snow-laden forests of North America, in the isolated lumber camps of Maine. In the late 19th century, physicians had begun hearing strange reports: French Canadian loggers who responded to sudden noises or commands with violent, uncontrollable reactions. The American neurologist George Miller Beard had been the first to study them in depth. He described visiting these logging camps, where the men lived and worked in close quarters, often under harsh conditions. Their peers often teased them mercilessly, using their exaggerated reflexes for amusement.

Their reflexes were startling. A slammed door, gunfire or a sharp order like "Jump!" could send them leaping into the air, flinging their tools or striking those nearby—seemingly unable to resist. The most remarkable detail, Beard noted, was not just the exaggerated physical response, but the instantaneous obedience. They obeyed without thought, as if their will had been momentarily overridden.

Meanwhile, in the icy reaches of Siberia, American naval officers stationed in the region reported a similarly unusual behaviour. Ship stewards displayed intense startle responses and a strange compulsion to mimic—repeating words, mirroring gestures, even echoing intonations. This condition, known as *myriachit* (meer-ya-kit), did not involve the same involuntary obedience seen in Maine, but the mimicry was uncanny. Russian psychiatrist Ardalion Tokarski documented it in detail, describing what he called "pathological imitation." He theorised it stemmed from an overactive reflex arc—a kind of reflexive epilepsy—though one with an unsettling degree of apparent volition. As in Maine, those afflicted were often mocked or provoked for entertainment.

Further south, in Southeast Asia, another pattern emerged. Among Malay communities, Western travellers observed a condition called *latah* (lah-tah), which bore a striking resemblance to the other disorders. Startled individuals would mimic words or actions with startling fidelity, sometimes blurting out obscenities or performing exaggerated gestures. But unlike the clinical framing in Siberia or Maine, *latah* was woven into the cultural fabric. It was a recognised—even expected—social phenomenon, treated with a blend of tolerance and humour. People with *latah* were often goaded by others, their involuntary performances accepted as part of village life.

For Tourette, these global case studies were electrifying. He pored over Beard's vivid accounts of the lumber camps, Tokarski's clinical notes from Siberia, and anthropological observations from

Southeast Asia. Despite differences in setting and cultural framing, all three conditions shared key features: exaggerated startle reflexes, involuntary mimicry, and reactions triggered by sudden stimuli.

In an 1884 article, Tourette proposed that these phenomena were not isolated curiosities, but regional variations of a shared underlying mechanism. Yet, as Tourette continued his work, he noticed a critical difference between these conditions and the patients he observed at the *Salpêtrière*.

One afternoon, back at the *Salpêtrière*, he sat across from a middle-aged woman on the examination table. She was nervous, her eyes darting as she twisted a handkerchief in her lap. Her body jerked suddenly—a sharp, involuntary movement—followed by an outburst of words, loud and jarring. She apologised immediately, cheeks flushed, but the tics came again.

Unlike the startle disorders he had studied, her symptoms had no external trigger. They erupted from within—unpredictable, spontaneous, and terrifying.

"Can you describe what you feel when this happens?" Tourette asked gently.

The woman hesitated, then said, "It's like... my body has a mind of its own. I can't stop it. The words just come out. The movements... they happen before I know what I'm doing."

Tourette nodded, taking careful notes. He could see the frustration and distress in her eyes, the exhaustion of living with a body that refused to obey her. The difference was critical. In the *Jumping Frenchmen*, *myriachit*, and *latah*, symptoms were provoked—reactive. But with this patient, the tics emerged unbidden, as though the brain's command centre had been hijacked.

Over the weeks that followed, Tourette documented case after case. Tics. Sudden, involuntary movements. Inexplicable vocalisations. No pattern of stimulus, no predictable cause. Just chaos, erupting from within the nervous system.

When he presented his findings to Charcot, the elder physician listened in thoughtful silence. At first, he debated whether this was simply another manifestation of the hyperstartle conditions. But as Tourette laid out his evidence—the spontaneous, persistent nature of these tics—Charcot began to see the distinction.

This was something new.

In time, Charcot fully endorsed Tourette's conclusions. The distinction he had drawn—between reflexive, culturally framed hyperstartle syndromes and the spontaneous, unpredictable tics seen in his patients—marked a critical shift in neurological thinking. It created a framework for separating superficially similar behaviours into distinct clinical categories, based on origin, not appearance.

By 1885, Georges Gilles de la Tourette had documented nine cases in painstaking detail and published his landmark study. He called the condition *maladie des tics*, or "disease of tics."

In recognition of his contribution, Charcot honoured him in a rare and public way. He named the condition after his protégé: *Gilles de la Tourette's illness*.

Now known as *Tourette Syndrome*.

Tourette was a curious figure—brilliant, erratic, and unforgettable.

His voice was rough and hoarse, constantly teetering on the verge of cracking. His gestures were abrupt, almost theatrical, as if his words alone could not contain his thoughts. His mannerisms drew attention wherever he went—part eccentric, part showman.

Fellow student Léon Daudet, once described him as "ugly, with a face like a Papuan idol with bundles of hair stuck on it." Tourette himself seemed a patchwork of contradictions, neither entirely kind nor cruel, disciplined nor chaotic, brilliant nor absurd. His personality was a kaleidoscope, shifting unpredictably. He could be argumentative, passionate, exasperating—often all in the same sentence. He was not a man who coasted. He collided.

This unpredictability often spilt into his work. Conversations with Tourette could veer off in a dozen directions, leaving his colleagues struggling to keep up. At first, his intensity amused his colleagues. Over time, it could wear them thin.

During one particularly heated debate, Tourette, his face flushed and his eyes blazing, argued passionately that general paresis (a brain condition causing dementia and paralysis) had no connection to syphilis. When colleagues pointed to mounting evidence, he would leap from his chair and exclaim, "It's my very firm idea!" over and over, pacing theatrically. His passion was undeniable. So was his impatience.

He brought this same flamboyance to student examinations.

"What disease begins with bleeding from the left nostril?" he once asked a candidate.

The student confidently replied, "Typhoid fever, sir."

Tourette paused, shook his head solemnly, and after a suspenseful silence, declared, "It's typhoid. You have scored no marks. You have failed."

On another occasion, Tourette's flair for the absurd took centre stage.

"Who are the three greatest doctors of the 19th century?" he asked a student. The response, "Laennec, Duchenne de Boulogne, and Charcot," was met with a dismissive wave.

"No, no. You don't know. They were my grandfather, my father, and me, *Coco!* That's why." Then, placing his hat on the bewildered student's head, he added, "That's why they'll erect a statue of me— made entirely of potassium bromide!"

And yet, beneath the absurdity, Tourette's intellectual contributions were serious—and substantial.

He published prolifically, writing on hysteria, neurasthenia (chronic fatigue and weakness), syphilitic myelitis (inflammation of the spinal cord due to syphilis), and forensic applications of hypnotism. His lectures were wide-ranging and ahead of their time, even attracting the attention of Sigmund Freud, who likely drew inspiration from his lectures.

He also championed unconventional therapies. Suspension therapy—where patients were hoisted in harnesses to relieve spinal pressure—and vibration therapy, using rhythmic mechanical pulses to relieve pain and improve circulation, were among his innovations. While many colleagues scoffed, Tourette's enthusiasm often proved infectious. Charcot, too, admired his drive and frequently lent support to his projects, helping him navigate the bureaucracy of academic medicine.

Tourette's life at the *Salpêtrière* was a whirlwind of energy. He juggled demanding clinical work with tireless research, often staying up late into the night. He also wrote under the pseudonym *Paracelsus*, penning biting satirical essays for *La Revue Hebdomadaire*, wielding his pen with as much precision as his scalpel.

But even the most tireless minds can falter.

In 1893, tragedy struck.

In 1893, Tourette's unstoppable momentum collided with catastrophe.

His young son, Jean, died of meningitis—a swift and brutal illness that tore through the protective membranes of the brain and spinal cord. The loss shattered him. Not long after, his mentor and closest ally, Jean-Martin Charcot, passed away.

The double blow was devastating.

Tourette buried himself in work, returning to the familiar confines of his consulting room as if order and duty might anchor him. But his focus faltered. The clinical notes became scattered. His writing lost its precision. The joy he once drew from medicine seemed to slip through his fingers.

Then, on a quiet afternoon, the fabric of his world tore open once more.

He was seated at his desk, reviewing a case file, when the door burst open. A woman stormed in—her eyes wild, her movements erratic, her face twisted with fury.

Tourette looked up, startled. He recognised the signs immediately: psychosis. Delusional thinking. She was not in control of herself.

"You!" she screamed. "You did this to me—you and your damned hypnotists!"

Before he could speak, she pulled a pistol from her coat and fired.

The bullet struck him in the neck. He fell to the floor, hands clutching the wound as warm blood pulsed between his fingers. Pain radiated outward, sharp and nauseating. His mind reeled, not just from the physical trauma, but from the disbelief—the sheer impossibility of it.

Tourette survived the attack, but something inside him did not.

The press seized upon the story. Headlines splashed across papers. The woman's claims—that she had been hypnotised against her will by one of Tourette's colleagues—ignited a public firestorm. The *Nancy School*, led by Hippolyte Bernheim, pounced on the opportunity. They argued that hypnotism could indeed override free will and provoke criminal behaviour—a direct challenge to the position Tourette and Charcot had long defended.

Suddenly, Tourette was at the centre of a bitter ideological battle.

The incident tarnished his reputation, casting a long shadow over his work in hypnotism and forensic psychiatry. Though his physical wounds healed, the psychological toll was undeniable. He became withdrawn, sombre, plagued by a sense of betrayal—by the public, by medicine, perhaps even by his own mind.

In 1894, with the help of his old colleague Brouardel, Tourette secured a new post: Professor of Legal Medicine at *St. Antoine Hospital*. He shifted his focus towards forensic work, but the damage had been done. The grief from his personal losses and the lingering trauma of the shooting began to weigh heavily on him. Tourette started to display symptoms of severe depression, and his mental health noticeably declined.

His once boundless energy gave way to fatigue. His sharp wit dulled. The grief that had long simmered beneath the surface began to show in erratic behaviour, melancholic silences, and bursts of agitation that unsettled colleagues.

In 1901, during a lecture, he collapsed mid-sentence. It marked a turning point—the moment his private struggle could no longer be hidden.

His mental state deteriorated rapidly. He became increasingly paranoid and delusional. His thoughts wandered. His lectures lost coherence. Friends and colleagues watched helplessly as the once-brilliant mind that had illuminated the halls of the *Salpêtrière* began to unravel.

Jean-Baptiste Charcot, son of his late mentor, recognised the severity of the situation and persuaded him to travel to Switzerland for what he described as rest and recuperation. Tourette agreed, unaware that he was being committed to a psychiatric hospital in Vaud.

There, doctors diagnosed tertiary syphilis—the final, devastating stage of the disease. It had been silently eroding his brain for years, explaining the disintegration of both body and mind. The seizures, the paranoia, the erratic behaviour—all now made sense, though the diagnosis brought no comfort.

Over the next three years, Tourette's condition worsened. Seizures became more frequent. Psychosis deepened. The sharp, curious mind that had once chased neurological mysteries across continents was gradually consumed by dementia.

His family remained by his side, offering quiet companionship as the man they knew disappeared in pieces. On 22nd May 1904, at the age of 46, Georges Gilles de la Tourette passed away, his once brilliant mind consumed by the ravages of neurosyphilis.

What is Tourette syndrome?

Tourette syndrome is a neurological condition that causes individuals to experience involuntary movements or sounds, known as tics. These tics can include blinking, shrugging, throat clearing, or repeating certain words. Tics are typically divided into motor tics (physical movements) and vocal tics (sounds or words). Symptoms often begin in childhood and can range from mild to more noticeable. While *Tourette syndrome* is a lifelong condition, many people experience an improvement in symptoms as they grow older. Treatments are available to help manage tics, allowing individuals to live whole and fulfilling lives.

CHAPTER 4

Irish Charm

THE BRIG, A sturdy two-masted sailing ship, heaved and groaned through the Mediterranean's angry waves. Every timber seemed to protest as the storm battered its hull. Above deck, the sky churned with black clouds, and the wind screamed like a living thing, driving rain sideways across the splintering wood. The ship, bound for Sicily, fought to stay upright, but it was clear to all aboard that the storm had the upper hand.

Below, in the dim cabin, Robert Graves lay curled on a narrow cot, his body wracked with pain from a stubborn illness that seemed intent on draining what little strength he had left. Sleep was out of the question; his mind raced with memories he would rather forget.

Austria. The dank, stinking dungeon where he'd spent ten days imprisoned as a suspected German spy. Ten days of accusations and indignities he'd never imagined he'd have to endure. It wasn't the physical suffering that haunted him most—though it had been brutal enough. It was the senselessness of it. The injustice gnawed at him still.

The ship lurched violently, sending a bucket skittering across the floor, and Graves blinked back to the present. One other passenger had shared the cramped cabin with him earlier in the voyage—a poor Spaniard, his unlikely companion on this ill-fated brig. They had forged a quiet camaraderie, united by circumstance, but now Graves was alone, the room thick with damp and dread. He wondered what fate awaited them in Sicily—assuming they survived the storm.

The door burst open, slamming against the wall. Graves' thoughts were cut short by the Spaniard's sudden entrance, his face pale as the foaming sea outside. "The crew," he stammered, "are abandoning the ship! They say we are to be left behind!"

Graves sat bolt upright, ignoring the sharp protest of his ribs. "What?" he barked.

"They're taking the lifeboat," the Spaniard said, his voice rising. "They don't think the brig can make it."

Graves didn't wait for more. He pulled on his cloak with quick, jerky movements and scanned the cabin. His gaze landed on an axe lying in the corner, discarded as if fate had placed it there for him. Without a word, he grabbed it and tucked it under his cloak before striding out into the gale.

The scene above deck was chaos. Rain lashed the desperate crew as they wrestled with the ropes securing the lifeboat, their faces twisted with fear. The captain bellowed orders, though his words were nearly lost to the roar of wind and waves. The brig was a wreck—its sails shredded, its pumps useless, the sea rising to engulf it.

Graves approached, his heart pounding.

"Stop!" he shouted, his voice cutting through the storm like the crack of a whip. The men froze, turning toward him with expressions of disbelief. "That boat will not survive these seas. Abandoning ship is madness!"

The captain spun to face him, eyes blazing. "This is none of your concern!" he snarled. "You and your companion will stay here!"

Graves felt a surge of defiance. "If we are to be left behind, then let us all be drowned together. It is a pity to part good company."

With that, he pulled the axe from beneath his cloak and, with a single swing, splintered the lifeboat's side. The wood cracked and buckled under the force of the blow.

For a moment, the captain seemed ready to attack, his hand going to the dagger at his belt. But something in Graves' stance—the calm conviction in his eyes—made him hesitate. The crew, witnessing their captain's uncertainty, faltered.

Graves seized the moment. "We can save this vessel," he declared. "But we must act quickly. The pumps have failed, but we can repair them."

For a heartbeat, no one moved. Then, as if shaken awake, the crew began to stir. Graves demanded action, his tone brooking no argument. They repaired the clogged pumps with leather strips from his boots and worked through the night, bailing water and patching the brig's battered hull. Slowly, the ship began to respond, rising with the waves instead of being crushed by them.

Hours passed in desperate labour, but under Graves' command, they managed to regain control. The leak was staunched, and the brig—battered but intact—rode out the worst of the gale. By dawn, the seas had calmed, and the first rays of sunlight broke through the clouds, casting a golden glow over the water.

Exhausted, drenched, and sore in every muscle, Graves stood at the helm. The crew, who had been on the verge of mutiny just hours earlier, now moved around him quietly. They cast glances his way—not with defiance, but with a newfound respect. Perhaps even gratitude. He had kept them alive when all seemed lost.

As the brig limped towards Sicily, the men whispered among themselves, stealing glances at the stranger who had seized an axe and assumed command in the storm. He was no sailor, that much was clear. But something in his manner, the certainty in his voice, had pulled them back from the brink. To them, it was almost miraculous, what he had achieved.

ᚼ

Robert James Graves was born in 1796 into a family steeped in academia, the eighth child of Richard Graves, the Dean of Ardagh.

Their household was one where books lined the walls, ideas were shared over meals, and curiosity was encouraged as both a virtue and a joy. Growing up amidst Ireland's rolling green hills, Graves absorbed the rhythms of nature and the intellectual traditions of his family, shaping the restless mind and vibrant spirit that would define him.

Even as a child, Robert Graves stood apart. He had a boundless energy that carried him through the wild landscapes of Ireland and the corridors of learning. His tutor—an exceptional man who would one day be remembered as the uncle of Oscar Wilde—quickly recognised the boy's sharp intellect and natural passion for the arts. Under his guidance, Graves became a lover of literature, a painter of bold, if imperfect, landscapes, and a prodigy in languages. His talent with languages was extraordinary, as though he had an instinct for untangling the mysteries tucked inside words.

At school in Downpatrick and later in Dublin, Graves excelled with effortless charm. By the time he entered *Trinity College* in 1811, he was already known for his ability to quote the classics with flair. Latin and Greek seemed as natural to him as his Irish brogue, giving even his scholarly recitations an air of poetry. His professors were astounded by his intellect; his peers were swept along by his wit and enthusiasm. Awards followed quickly—none more prestigious than the Gold Medal, the crowning achievement of his academic career.

But while the ancient texts stirred his imagination, Graves found himself increasingly drawn to medicine. It wasn't just the science of it that intrigued him, but the human side. Perhaps it was his desire to connect with people, to understand suffering and healing as intimately as he understood words. By 1818, medical degree in hand, he was ready to leave Ireland in search of deeper knowledge.

Graves' wanderlust took him across Europe, through the great intellectual capitals of the age—London, Edinburgh, Berlin, Vienna, Copenhagen, Paris. Wherever he went, he threw himself in completely, absorbing new ideas and building friendships with some of the leading minds of the era. These journeys weren't just academic—they were transformative. Graves didn't simply collect facts and techniques; he embraced cultures, challenged assumptions, widened his horizons. And in doing so, he inspired others to do the same.

Robert Graves was impossible to ignore. Tall and broad-shouldered, with dark hair that refused to be tamed, he moved through the world with a kind of quiet confidence that invited trust. His skin, darkened by sun and travel, gave him a striking, weathered look, while his face— always shifting, always alive—flashed easily with humour or reflection. His eyes, sharp and full of fire, seemed to miss nothing.

And then there was his voice—deep and warm, edged with the music of his Irish accent. It wasn't just what he said that drew people in, but how he said it. Even the most complex ideas seemed to become clear and close when he spoke, like a story you were being let in on. Whether in a crowded lecture hall or over a quiet table, Graves made people feel seen, drawn in, and energised by the world around them.

One of the most harrowing episodes of Robert Graves' travels began on a cold, misty morning in the Austrian countryside. The quiet beauty of the landscape had lulled him into a rare moment of peace when a sharp tug on his arm brought him to a halt. Turning, he found himself surrounded by Austrian soldiers, their faces set with suspicion. The leader barked an order, and before Graves could fully grasp what was happening, he was told he was under arrest— suspected of being a spy.

For a moment, Graves was too stunned to speak. When he did gather himself, he explained, in fluent German, that he was merely a traveller, passing through on foot. But the absence of a passport raised alarm, and his command of their language only made things worse. To them, the idea that an Irishman could speak German so precisely was improbable. Perhaps even threatening. Though his tone remained calm and logical, their suspicion only deepened with each word.

He was led to a dark dungeon, its air thick with damp and the slow decay of forgotten things. Cold crept into his bones. The food, when it came, was barely edible, and the silence between the guards was broken only by their scornful remarks. The worst part wasn't the hunger or the cold, or even the solitude. It was the disbelief—being treated not as a man but a deception, a lie. For ten long days, Graves endured that cell, his frustration sharpening into something more lasting. Yet even in the thick of that injustice, his spirit held. When he was finally released, there was no apology—only a curt dismissal. He emerged thinner, weaker, but quietly defiant.

Not long after, Graves boarded a weathered brig bound for Sicily, crewed by Sicilians and carrying just one other passenger: the poor Spaniard who would become his companion in adversity. Their friendship, formed in close quarters and strange silence, was a small comfort on a voyage that would soon veer into calamity.

A storm engulfed the brig, violent and furious. Waves pounded the hull, the sails were torn to ribbons, and the crew, gripped by panic, prepared to flee. Graves, still weakened from his imprisonment, found himself thrust into a crisis. But something in him rose to meet it. He acted—seizing an axe, rallying the crew, and taking charge when no one else could. Through sheer force of will, ingenuity, and a refusal to abandon hope, he led them through the night, patching the ship, unclogging the pumps, keeping them afloat when the sea tried to claim them.

By dawn, the storm had passed, and the brig limped onward. The crew, who had nearly left him behind, now watched him with something close to reverence. He had kept them alive. And though he was not a sailor, he had done something few captains could.

But Graves was never one to linger in triumph. His gaze was always moving forward, curious, hungry for meaning. Even in the wake of storms, he returned to beauty. While crossing the Swiss Alps not long after, he met the renowned painter J.M.W. Turner. The two struck up an unlikely friendship—Turner, the restless master of light and movement, Graves, the doctor with a poet's eye. They travelled together for months, sketching scenes, chasing mountains and sunrises, trading thoughts on art and life. Turner's paintings spoke to Graves in a language he understood instinctively: of transience, power, and wonder. Their time together—from the jagged peaks of the Alps to their parting in Rome—became one of Graves' most treasured memories. A perfect meeting of art, friendship, and discovery.

Of course, Graves' travels weren't all art and adventure. In Berlin, he studied under Christoph Wilhelm Hufeland—one of Europe's leading physicians, known for his pioneering work on infectious diseases. Graves was captivated. Hufeland's lectures crackled with insight, and his clinical methods demanded observation, not just theory. Graves threw himself into the work with his usual intensity, taking notes, asking questions, testing ideas. He didn't just want to understand medicine—he wanted to live it.

That knowledge would be tested sooner than he expected. When Graves returned to Ireland during a typhoid epidemic, the lessons he'd learnt abroad became tools in the fight to save lives. He wasn't content to simply diagnose and treat. He wanted to shift how medicine itself was taught and practised.

By the time he set foot on Irish soil again after three years away, he was changed. Every storm, every betrayal, every friendship had shaped him. The wide-eyed student who had once wandered the continent was now something more: a man sharpened by hardship, deepened by beauty, and driven by an unshakable belief that knowledge should be used—urgently, fearlessly—to make the world better.

When Graves stepped off the ship and into the familiar streets of Dublin, a wave of relief and quiet anticipation washed over him. The air carried that familiar tang of salt, and the city, with its muddle of conversation and cobbled streets, felt unchanged. But he knew he wasn't the same man who had left. He had seen what was possible in Europe—the sharp edge of medical thinking, the boldness of change—and it burned in him now, eager for a place to take root. Still, a flicker of doubt lingered. Would his ideas find a home here? Or would they face the resistance of tradition?

The answer came quickly. His return in 1821 caused a ripple through Dublin's tight knit medical community. Stories spread of the young doctor who had studied with Hufeland, survived an Austrian prison, saved a ship in a storm, and sketched with Turner in the Alps. Some were impressed. Others were wary. But Graves didn't care for reputation. Ireland needed change, and he meant to be part of it.

It wasn't long before he joined the staff at the *Meath Hospital*, a once-promising institution now in slow decline. One morning, sunlight streamed through the tall windows of a ward as a group of physicians gathered at a patient's bedside, their voices low and speculative. Students loitered at the back, craning their necks to catch a glimpse or overhear a scrap of diagnosis, but unable to see or hear much. Graves watched them from across the room. He remembered being one of them once, shut out, hungry for real learning but kept at a distance.

Enough, he thought.

He stepped forward, his voice rising above the murmured conversation.

"Ladies and gentlemen," he began, his tone firm but not unkind, "a medical education is not to be gained merely by listening to lectures or reading books. From the very commencement, the student should set out to witness the progress and effects of sickness—and ought to persevere in the daily observation of disease during the whole period of his studies."

The room stilled. The physicians fell silent. The students looked up.

Graves gestured for them to come closer.

They hesitated—surely this was a breach of protocol. But his nod was clear, encouraging. Slowly, awkwardly, they stepped forward. Under his guidance, they examined the patient, discussed the symptoms, fumbled through tentative diagnoses. At first they were nervous, unsure of themselves. But Graves was patient, asking questions, drawing them out. By the end of the session, they were no longer spectators. They were learning.

Graves' approach was revolutionary—and not without its critics. Some colleagues whispered that he was reckless, that he blurred the line between teaching and theatre. But Graves knew what he was doing. Medicine, to him, wasn't an abstract science. It was human. You had to see. To listen. To be present.

During a rare quiet moment in the hospital, he confided as much to his friend and fellow physician, William Stokes.

"How else will they learn?" he said, almost to himself. "We owe it to them—and to our patients—to teach by showing."

Graves' influence stretched beyond the walls of *Meath Hospital*. His belief in close observation and human connection wasn't just reserved for the lecture room—it shaped how he treated his patients. And during the typhoid epidemic that swept through Dublin, that philosophy was put to the test.

At the time, standard medical thinking dictated that fever patients should be starved. Food, it was believed, would fuel the illness. But Graves didn't accept that. He'd seen too many patients waste away under this regimen, their bodies fighting battles with no strength left to spare. Drawing on the clinical insights he'd gathered across Europe—and his own instincts—he began to do something radical.

He fed them.

Soup. Broth. Simple meals in careful portions. It went against everything his colleagues had been taught. The debates were heated. Some dismissed his methods as dangerous. But the results were hard to ignore. His patients began to recover. Slowly, yes—but steadily. Graves didn't claim to have all the answers, but he trusted what he saw. He listened to the body, not just the books.

Decades later, during the devastation of the Irish Potato Famine, his ideas would find even deeper resonance. Faced with overwhelming loss—over a million lives—Graves' patient-centred approach offered a measure of hope. He couldn't solve the famine. But he could do what he had always done: feed, care, and treat.

Even in the midst of suffering, Graves never lost his sense of humour. It wasn't a performance or a distraction—it was part of how he stayed grounded, how he reminded himself (and others) that medicine, at its heart, was about people.

One morning, as he and Stokes moved steadily through the wards, they approached the bed of a fever-stricken man. The patient's face was pale, his breathing shallow. Graves placed two fingers gently on his wrist, his brow furrowing as he felt the faint, uneven pulse.

He stood there in silence for a moment, then turned to Stokes, his tone dry but warm.

"You know, William," he said, "this man needs food, not starvation. Half the treatments in the textbooks wouldn't do him as much good as a bowl of soup."

Stokes chuckled, the tension in his shoulders easing.

"You're probably right," he replied. "You usually are."

They shared a look, a quiet moment between two men who had seen more than most. Graves reassured the patient gently, prescribed small meals and rest, then moved on.

But just before they reached the next bed, Graves paused mid-step, a spark of mischief in his eye.

"William," he said, glancing over his shoulder, "lest when I'm gone you find yourself at a loss for an epitaph, let me save you the trouble."

Stokes raised an eyebrow, curious.

"Oh?"

Graves smiled—light, irreverent, but with that unmistakable note of something truer beneath.

"Simply this: *He Fed Fevers.*"

Stokes laughed, shaking his head. "I'll make a note of that, Robert," he said. "Though I hope I won't need it anytime soon."

For Graves, humour was never an escape. It was a form of presence—a way of keeping his feet on the ground and his eyes on the people who needed him most. In the face of illness, bureaucracy, and grief, it reminded him of why he chose medicine in the first place.

Graves' rise within the medical community was steady and hard-earned. He was appointed Professor to the Institutes of Medicine in the Irish College of Physicians, and he published widely on medical and physiological subjects. Titles and honours followed: President of the Royal College of Physicians of Ireland, Fellow of the Royal Society of London. Yet he never lost touch with the ward floor, or with the students who followed him from bedside to lecture hall, hungry for the kind of learning only he seemed able to provide.

He believed in a dynamic medicine rooted in observation, in curiosity, in direct human contact. His students didn't know him as a distant figure delivering pronouncements from the front of a room. He walked among them, challenged them, encouraged them. His mantra was simple but powerful: *"Learn the duty as well as taste the pleasure of original work."* It was more than advice. It was a way of thinking, one that shaped the *Dublin School of Medicine* into an institution respected across Europe.

One morning at *Sir Patrick Dun's Hospital*, Graves was preparing for rounds when a colleague approached him with a note of unease.

"Professor Graves," the man said, lowering his voice, "there's a patient presenting with unusual symptoms. No diagnosis seems to fit."

Graves paused. These were the kinds of cases that stirred something in him. They weren't just puzzles—they were opportunities.

He folded his notes and walked toward the ward, not knowing that this patient would lead him to the most enduring discovery of his career.

He folded his notes and walked toward the ward, not knowing that this patient would lead him to the most enduring discovery of his career.

At the bedside sat a young woman, visibly anxious. Her fingers twitched where they gripped the blanket, and her breath came quick and shallow. Graves' gaze lingered on her eyes—they protruded so markedly that her lids no longer fully closed, even at rest. A bulge at her neck revealed a swollen thyroid. Something was clearly wrong.

Graves offered a quiet smile as he approached.

"Miss, I'd like to examine you more closely, if I may," he said gently.

She nodded; her wide eyes fixed on him. He took her wrist and timed her pulse—rapid, irregular.

"Your pulse is very quick," he said softly, then reached to palpate her neck. "Do you often feel nervous or agitated?"

"All the time, doctor," she replied, voice trembling. "Even when there's no reason for it."

He made a note in his journal, then leaned closer, lowering his voice.

"And your eyes," he asked, "have they always appeared this way? Or is it something new?"

"It started a few months ago," she whispered, looking away. "But it's getting worse. I can barely close them now."

Graves nodded. He kept his tone calm, but his mind was already fitting the pieces together. A rapid pulse. Anxiety. A visibly enlarged thyroid. Bulging eyes. It wasn't just a coincidence—it was a pattern.

Over the next few weeks, two more cases appeared, uncannily similar. One was brought to him by William Stokes, who stopped him in a corridor, visibly animated.

"Robert," he said, "I've just seen another one. Eyes, thyroid, palpitations—the same. It's remarkable, isn't it?"

Graves raised an eyebrow, intrigued. "The consistency is extraordinary," he said. "I think we may be seeing a distinct condition."

He and Stokes pored over their notes, compared cases, debated possible explanations. But Graves, ever the observer, trusted what he saw. These weren't scattered symptoms. They were signs of something real. And they needed to be understood.

In 1835, he published his observations in the *London Medical Society Clinical Lectures*, describing in vivid detail the combination of rapid heart rate, thyroid enlargement, and exophthalmos—the protruding eyes. At *Meath Hospital*, he lectured to packed rooms.

"Ladies and gentlemen," he began, pointing to his sketches, "observe the swelling of the thyroid gland in these patients. Note the rapid pulse. The nervousness. And, most remarkably, the protrusion of the eyes. These symptoms together form a distinct clinical picture."

His voice was measured, but the air was electric. Students scribbled notes. Colleagues nodded, murmuring in agreement. Graves' clarity of vision—and the certainty of his tone—left no doubt that something important had been uncovered.

He wasn't the first to notice the condition. Caleb Hillier Parry had described similar symptoms decades earlier, though his work was only published after his death. Carl von Basedow in Germany had also written on the subject. In fact, in parts of the world, the disease still bears Basedow's name. But it was Graves' name that stuck. His vivid descriptions, his standing in the medical world, and his broad international connections helped solidify the link.

A key influence came from France, where physician Armand Trousseau—an admirer of Graves—popularised the term *"Graves' disease"* in his lectures and publications. The name took hold.

But for Graves, naming it wasn't the point. Understanding it was. This discovery, like everything else in his career, began not with theory, but with a patient. A human being. A story unfolding in front of him.

Though Graves was known for his brilliance, charisma, and restless energy, his personal life was marked by deep heartbreak—losses that shaped him in ways no professional challenge ever could.

In 1821, shortly after returning to Dublin, he married Matilda Jane Eustace. Their union brought joy and stability, but the happiness was short-lived. Within four years, tragedy struck. Matilda died during childbirth, and their infant daughter, Eliza, passed away soon after. The double blow devastated Graves. Grief settled over him like winter frost—quiet, numbing, and inescapable.

In 1826, seeking comfort and companionship, he married again—this time to Sarah Jane Brinkley. For a moment, there was hope of healing. But fate had other plans. Within a year, Sarah too died in childbirth, along with their newborn daughter. Another light extinguished. Another grave to tend. The weight of repeated loss could have hollowed him out. Instead, Graves buried himself in work, pouring his grief into purpose.

Through it all, he remained devoted to medicine, to teaching, to building something larger than himself. And eventually, life offered a measure of peace. In 1830, he married Anna Grogan. Where earlier chapters of his life had been marked by drama and brilliance, this was something steadier. Anna brought warmth and quiet strength. Together, they had six children. She would outlive him by two decades, and her presence offered him the family legacy he had once feared he would never have.

But even as his family life found balance, the demands of his career continued to take their toll. By the early 1840s, the signs of exhaustion were difficult to ignore. The tireless energy that had once defined him began to fade. In 1841, he resigned his chair of medicine. Two years later, he stepped down from his role at Meath Hospital, though he accepted the presidency of the Royal College of Physicians of Ireland. His colleagues still respected him. His students still spoke of him with awe. But something had shifted.

Portraits from those years reveal the change clearly. The animated features were heavier now. The glint in his eyes—once sharp with mischief, conviction, and curiosity—had dulled. He looked older than his years, not just in body, but in spirit.

Graves died of liver cancer on 20th March 1853 at his country residence, aged just 56. It was a quiet end to a life that had burned so brightly for so long. But the light he left behind did not go out.

His insistence on bedside teaching, on active, human-centred medicine, changed how doctors learned and how patients were treated. He challenged orthodoxy, questioned harmful conventions, and taught others to do the same. He didn't just teach medicine. He transformed it.

What is Graves' disease?

Graves' disease is an autoimmune disorder that causes the thyroid gland, a small, butterfly-shaped gland in the neck that regulates metabolism, to become overactive. This leads to an excess of thyroid hormones (a condition called hyperthyroidism). People with *Graves' disease* often experience symptoms such as a rapid heartbeat, weight loss, anxiety, tremors, and excessive sweating. A distinctive feature of the disease is the bulging of the eyes (exophthalmos), along with enlargement of the thyroid gland, known as a goitre. Even today, *Graves' disease* is a significant focus of research and treatment in the field of endocrinology, the study of hormones and glands.

CHAPTER 5

A Mother to All

IN THE HISTORY of medicine, eponyms tell a particular kind of story. They recall a time when the field was almost entirely dominated by men, their names etched into textbooks and whispered down hospital corridors like monuments to progress. But those names also cast long shadows. They obscure the voices of women who, for centuries, were kept at the margins—present, often essential, but rarely acknowledged. By the late 19th and early 20th centuries, when medical eponyms flourished, women were still fighting for their place in the room. And because of that, the list of names we remember remains one-sided.

Still, some women refused to let those barriers hold them back. They broke through—quietly, stubbornly, sometimes explosively—leaving behind legacies that reshaped medicine forever. Among them, Virginia Apgar stands not just as a name remembered, but as one celebrated. She didn't simply enter the profession. She rewrote its rules.

Apgar's story began in June 1909 in Westfield, New Jersey, in a home where curiosity and creativity were everyday currency. Her father, Charles, worked in insurance but lived for invention. His basement was a chaos of telescopes, radios, and half-built contraptions—a playground of precision and wonder. Young Virginia often hovered by his side, watching his hands, mimicking his movements.

"Pay attention," he'd tell her. "You have to see what others don't."

Her mother, Helen, filled the house with music. Virginia played the violin, her brother the piano, and their evenings often unfolded in a swirl of melodies and laughter. Music wasn't a pastime—it was a thread that stitched the family together. Even years later, when life became complicated, Apgar would return to her violin, letting its familiar notes carry her somewhere quieter, more certain.

But life was not always kind. Her eldest brother died of tuberculosis. Her other brother struggled with chronic illness. Loss arrived early and taught her something lasting: life is fragile, and every effort to preserve it matters.

By the time she graduated high school in 1925, Apgar knew where she was headed. Medicine. At *Mount Holyoke College*, she threw herself into zoology, physiology, and chemistry. Money was tight, and she picked up odd jobs—whatever would cover the next term's tuition. One task involved catching stray cats for physiology labs. Unlikely work, unglamorous too, but she did it with the same grit she applied to everything else.

In 1929, she entered *Columbia University College of Physicians and Surgeons*. Medicine wasn't a career for her; it was a calling. Surgery, especially, lit something in her. It was tactile, precise, inventive. She excelled. Her mentors noticed her hands—steady, skilled—and her instincts, sharp as a scalpel. But when she graduated in 1933, reality hit hard.

Dr Allen Whipple, one of her surgical mentors, called her into his office.

"Virginia," he said gently, "you've got the makings of a great surgeon. But the field isn't ready for you."

It wasn't a dismissal. It was a warning. Whipple had seen too many gifted women ground down by a profession that refused to yield. He didn't want that for her.

The words stung. That night, Apgar scribbled furiously in her diary. *Is this it? Is this where I give up?* But giving up wasn't her way. Whipple's advice stayed with her, not as a dead end, but as a signpost. He had quietly suggested another path: anaesthesiology (the medical speciality focused on the use of drugs and techniques to prevent pain and provide sedation during surgery and other procedures). At the time, it barely counted as a medical speciality. It was poorly defined, often overlooked. At first, it felt like a consolation prize. But the more she considered it, the more it intrigued her. It wasn't established. It wasn't crowded. It had no gatekeepers.

It was a field she could build from the ground up.

So, she took the leap. Apgar trained under Ralph Waters at the *University of Wisconsin*, and later under Emery Rovenstine at *Bellevue Hospital* in New York. These were men shaping the frontier of a new discipline—and Virginia was the only woman among them.

A 1937 photograph of Waters and his residents captures the reality: 15 men in starched coats and quiet authority —and Apgar, standing among them, the lone woman in a profession that barely acknowledged her right to be there.

There was no housing for women in the programme, so she slept in her mentor's office. It was cold and lonely, but she managed. Later, she moved into a maid's quarters room, which felt like a luxury in comparison, though it couldn't compete with the camaraderie of her male colleagues. She imagined them sharing cigars and laughter in men-only clubs, forming the kind of easy bonds she was excluded from.

Her diary from those years reads like a ledger of frustrations: the barriers, the slights, the isolation. *I can't fight if I'm locked out of the room.* But she kept fighting anyway. She turned every closed door into fuel. Because once she entered the operating theatre, there was no question of whether she belonged. She was brilliant. She was steady. And she knew it.

In 1938, she returned to *Columbia-Presbyterian Medical Centre*— this time as director of its newly established anaesthesia division. She was the first woman to head a speciality division at the institution. It was a milestone, yes. But it also marked the start of a marathon.

Her role was everything at once: administrator, recruiter, teacher, clinician. She organised staffing rosters, taught courses, developed protocols—all while keeping one foot in the operating room. The demands were immense, but she thrived on the challenge.

Anaesthesiology still fought for legitimacy. Many surgeons resisted working alongside anaesthetists, dismissing the role as peripheral at best, an annoyance at worst. Some called it interference. But Apgar had a disarming presence—pragmatic, good-humoured, and unshakably patient. She didn't argue her way into acceptance. She earned it. Case by case. Conversation by conversation.

One by one, her colleagues came around. They saw what she saw: that anaesthesia wasn't a luxury—it was lifesaving. Her presence in the theatre didn't dilute their authority; it made their work safer, more precise, more humane.

Then came the war.

When World War II broke out, male physicians left for military service in droves. The anaesthesia division was gutted. What remained rested squarely on Apgar's shoulders. Long days bled into longer nights as she juggled clinical care, staff training, emergency cover, and recruitment. The pressure was ruthless.

There were moments—late at night, sitting alone in her office, surrounded by the ticking of clocks and the smell of old paper— when the exhaustion almost broke through. But it never did.

Because Apgar didn't stop. Not for weariness. Not for bitterness. And certainly not for anyone's permission.

When the war ended, a new chapter opened for anaesthesiology. Interest surged. The speciality was no longer seen as fringe; it was gaining respect, even momentum. Hospitals began formalising departments, and anaesthetists—long treated as assistants—were finally being seen as essential partners in care.

It should have been Apgar's moment.

She had steered the division through its toughest years. She had trained the staff, shaped the protocols, held everything together when it nearly fell apart. But when *Columbia-Presbyterian* created a formal Department of Anaesthesia, Apgar wasn't chosen to lead it. The post went to Emmanuel Papper, a man with impressive academic credentials and a polished research portfolio.

The decision hurt. She had worked in real-time crises, where the stakes were life and death—not theory and publication. Her approach was clinical, practical, patient focused. But academia valued papers over practice, titles over tenacity.

She likely sat with that injustice for a while. Wondered whether it would always be this way—that a woman's proof had to be twice as loud, twice as long, just to be half as heard.

But wallowing wasn't in her nature.

Instead, she shifted course. She accepted a faculty position at the *Columbia University College of Physicians and Surgeons*, becoming the first woman to hold a professorship there. If she wasn't going to lead in name, she would lead in action.

She turned her energy to building something bigger: formal residency programmes, consistent standards for training, better recognition for the field. She mentored a generation of doctors who had never known a time when anaesthesiology wasn't essential. Her work laid the foundation for how the speciality is taught and practised today.

Apgar didn't need a title to lead. She shaped the future of her field from the inside out.

Apgar's career took a decisive turn at the *Sloane Hospital for Women*, Columbia's maternity wing. It was here, surrounded by the rhythm of contractions, the clamour of delivery rooms, and the cries—or silences—of the newly born, that her attention began to shift.

She had always been attuned to patterns. And in obstetric

anaesthesia, the pattern that troubled her most was what happened in the minutes after birth—when, too often, nothing happened at all. She had long been troubled by the high rates of infant deaths within the first 24 hours after birth, a tragic reality that persisted despite advancements in other areas of medicine.

The mother, exhausted and vulnerable, was the focus of the team's attention. Rightly so—but not always fully. The baby, meanwhile, was left to fate. Most cried. Some didn't. Sometimes they breathed. Sometimes they didn't. And too often, those early moments—so critical, so fleeting—were lost to chance.

Apgar saw it over and over. A newborn's limp body, a pause too long before intervention, a missed sign of distress. No protocol. No standard. No urgency, unless it was obvious and already too late.

The silence of those seconds stayed with her. They followed her home, echoed in her thoughts at the end of a long shift. She began documenting what she saw. Colour. Tone. Breathing. Reflexes. She started to see the shape of something clearer than instinct—something measurable.

She had a unique vantage point. As an anaesthesiologist, she stood at the intersection of cause and effect—seeing not just what happened, but *why* it might have happened. She could observe both the procedure and its consequences, linking the two with precision and empathy.

In her mind, the question wasn't *if* something could be done. It was *how soon* she could start.

The moment came, as revolutionary moments sometimes do, without fanfare.

It was 1949, a typical morning in the *Columbia-Presbyterian* cafeteria. The clatter of trays, the hum of early conversations, and the sharp smell of coffee hung in the air. Apgar, as she often did, sat with students and residents, her sleeves rolled up, her presence equal parts commanding and approachable.

Across the table, a young medical student furrowed his brow, the way she had done a hundred times before him.

"Dr Apgar," he said, "I can't find a reliable way to evaluate newborns in the delivery room."

The question stopped her mid-sip.

It wasn't a new frustration—not to her. But there was something in the simplicity of the moment, the directness of the question, that brought it all into focus. The babies she'd seen. The silences that had haunted her. The missed seconds. The missing system.

She glanced around. Her eyes landed on a small card on the table— *"Please bus your own tray"*—and, without hesitation, flipped it over. She pulled a pen from her coat pocket and began to write. Quickly. Instinctively.

Five signs. The ones she'd learnt to read at a glance, after years in the delivery room:

- Heart rate
- Respiration
- Colour
- Muscle tone
- Reflex irritability

As she wrote, the cafeteria noise faded. Her mind was back in the delivery room—kneeling beside a silent infant, willing it to cry, to colour, to move. She wasn't just making a list. She was building a bridge between what was seen and what was done.

She held up the card.

"You'd do it like this."

The students leaned in. Something about the moment—the crispness of the idea, the elegance of its execution—made them fall silent.

Apgar didn't wait. She left the cafeteria and walked straight to the labour and delivery ward. The list in her hand. The idea already in motion.

What Virginia Apgar had scribbled on a tray card that morning was deceptively simple. But its brilliance lay precisely in that simplicity.

The *Apgar Score* gave clinicians a common language for urgency—a way to see, to assess, and to act. It was not diagnostic, but it was directive. A baby with a low score would no longer be overlooked. The score turned those crucial first minutes of life into structured, lifesaving moments.

Each newborn was assessed at one minute after birth. Five signs, each scored 0 to 2. The sum would provide a total between 0 and 10. Not a diagnosis—but a signal. A flag. A prompt to act, or reassurance to watch. And then something delightful happened: her surname became an acronym:

- **A**ppearance (colour) – Completely pink earned a 2; pink body with blue extremities, a 1; pale or blue all over, a 0.
- **P**ulse (heart rate) – Over 100 beats per minute scored a 2; under 100, a 1; no pulse, 0.
- **G**rimace (reflex irritability) – A vigorous cry or strong reaction got a 2; a weak response, a 1; no response at all, 0.
- **A**ctivity (muscle tone) – Active motion scored 2; some flexion, 1; limp or floppy, 0.
- **R**espiration (breathing effort) – Strong, steady crying earned 2; weak or irregular breaths, 1; no breathing, 0.

It was elegant, repeatable, and effective. Best of all—it could be used by anyone trained to observe—by obstetricians, midwives, nurses, paediatricians alike.

Still, not everyone welcomed the change. Some obstetricians bristled at the suggestion that someone should score *their* delivery. Would a low *Apgar* score reflect badly on them? Would they be judged?

Apgar, as always, offered a solution rather than a quarrel.

Let a neutral party—say, a circulating nurse—handle the scoring.

That way, the score was about the baby, not the doctor. It was classic Apgar: sidestep ego, get back to the work.

She refined the system with the help of a research nurse, tested it, and presented it to the *International Anaesthesia Research Society* in 1952. It was published the following year.

By the early 1960s, the *Apgar Score* had become standard practice in hospitals across the United States—and soon, across the world. It was adaptable, scalable, and suited to settings from world-class medical centres to remote field clinics.

And most importantly—it saved lives.

But Apgar wasn't finished.

Recognising distress wasn't enough. Action had to follow. She turned her attention to neonatal resuscitation, asking not just how to spot a struggling baby, but how best to intervene. Her research explored the effects of maternal anaesthesia on newborns, revealing that general anaesthetics often crossed the placental barrier (the organ that connects the developing baby to the mother) and left infants sluggish, sedated, slow to respond. Lower *Apgar* scores often followed.

She began championing the use of spinal anaesthesia in caesarean sections—keeping the mother awake while sparing the baby from unnecessary exposure to depressant drugs. Outcomes improved. Babies scored higher. Lives were changed.

Once again, Apgar had seen the pattern. And once again, she had acted.

By the late 1950s, Apgar had seen too many newborns born into struggle—tiny bodies burdened by conditions that might have been prevented. The delivery room had shown her what was possible, but it had also shown her the limits of clinical care. To make a real difference, she needed to move further upstream. Prevention, not just reaction. Policy, not just intervention.

In 1959, she made a bold decision. She left Columbia and enrolled at the *Johns Hopkins School of Hygiene and Public Health*, earning a Master of Public Health. It wasn't a retreat. It was a pivot. She was equipping herself to fight a larger battle.

Soon after, she joined the *March of Dimes Foundation*. What had once been a polio-focused organisation was shifting its mission toward preventing birth defects. Apgar arrived at exactly the right moment—with the credibility of a seasoned clinician and the fire of a reformer.

Her work there was intense. She travelled across the country, speaking to everyone from doctors and policymakers to new parents and community volunteers. She made the science understandable, the stakes personal. She believed in research, yes—but she also believed in raising awareness. If families understood the risks, if governments invested in prevention, then fewer babies would be born into lives of hardship. That was the goal.

One of her most urgent missions was the campaign for universal vaccination against rubella—German measles. At the time, the dangers were not widely known. But Apgar had seen the consequences up close.

Rubella infection during pregnancy could cause devastating harm: deafness, blindness, heart defects, intellectual disabilities—or all too often, miscarriage or stillbirth. During the 1964–65 rubella pandemic, the United States recorded over 12 million cases. 11,000 pregnancies ended in miscarriage or therapeutic abortion. 20,000 babies were born with congenital rubella syndrome.

Apgar knew these numbers weren't just statistics. They were futures derailed. Families broken. Losses that could have been avoided.

She became a tireless advocate for vaccination, speaking with urgency but never alarmism. She didn't scold. She explained. She persuaded. And people listened. Her voice—grounded, compassionate, determined—helped turn the tide.

At the *March of Dimes*, she also helped double the organisation's annual income. That wasn't a fluke. It was her energy, her clarity, her ability to connect with every kind of audience. The increased funding allowed the foundation to expand its research, its outreach, its impact. Her fingerprints were on all of it.

For Apgar, the logic was simple. Every child deserved the best start in life. If that meant leaving behind the comfort of clinical work, she did it. If it meant learning a new language of policy and public health, she learnt it. She wasn't chasing prestige. She was chasing outcomes.

For all her discipline and drive, Virginia Apgar never lost sight of the things that made her feel most alive. Chief among them was music.

From her earliest days, music had been a constant companion. She played the violin as a child, then later added cello. Her instruments travelled with her wherever she went—to work, cross-country tours, international conferences. They weren't luggage. They were lifelines. When days ran too long or the world felt too heavy, she would lose herself in the melodies, her bow gliding across the strings. Music didn't take her away from her work. It reminded her why she did it.

She played in orchestras and chamber groups whenever she could. *The Teaneck Symphony. The Amateur Music Players. The Catgut Acoustical Society.* Music wasn't just a hobby for Apgar; it was her sanctuary. She approached music with same precision and passion she brought to her work: with rigour, delight, and a kind of reverent mischief.

One afternoon in 1956, that mischief met its match.

Apgar was visiting a preoperative patient—Carleen Hutchings—a science teacher and musician. The hospital room was quiet except for the soft sound of sanding. Hutchings was bent over and was hand-carving a viola in her hospital bed, despite the looming prospect of surgery. Apgar paused at the door, intrigued by the delicate curves taking shape under Hutchings' tools.

"What are you working on?" she asked, stepping closer.

Hutchings looked up and smiled. "I'm building a viola. I've been studying how stringed instruments produce sound. Thought I'd try my hand at making one."

Apgar was captivated. As Hutchings spoke about acoustics, tonewoods, resonance and scrollwork, Apgar felt something spark. Here was a whole world she hadn't explored—where craftsmanship met physics, where sound could be shaped with your hands. It was as if Hutchings had opened a door to the exact intersection of Apgar's passions.

Soon, her nights changed. Midnight to 2:00 a.m. became her sacred hours. Her apartment transformed into a luthier's workshop. The smell of sawdust drifted into the hallways. Tools clinked gently across wood. Her neighbours, woken by the soft rasp of sanding or the faint thud of a chisel, became silent witnesses to her latest obsession.

She carved four instruments in total: a violin, a mezzo violin, a cello, and a viola. Her workspace overflowed with fine wood, tuning pegs, and half-assembled soundboards. Under the glow of a desk lamp, she shaped scrolls and smoothed maple, each curve as considered as a surgical incision.

But the best story came in 1957, when she rang Hutchings with barely contained excitement.

"Carleen," she said, "you won't believe what I found. There's a piece of curly maple being used as a shelf in one of the pay phone booths at the hospital. It's perfect for a viola back. The grain is gorgeous."

"Can you get it?" Hutchings asked.

"I've tried," Apgar replied. "But the admin office won't budge. Bureaucracy."

A pause.

"So what should we do?" Hutchings asked, a smile already forming.

Apgar's eyes twinkled with excitement as an idea formed in her mind. "What if we just... liberate it?"

Hutchings chuckled, sensing an adventure. "Liberate it?"

"Meet me at the hospital's ambulance entrance at the *Harkness Pavilion at Columbia-Presbyterian Medical Centre* around midnight," Apgar suggested, her excitement growing.

"Alright, I'll be there," Hutchings agreed, her voice tinged with anticipation.

On a cold January night, the two women met at the ambulance entrance of the *Harkness Pavilion*. Apgar donned her white doctor's coat, the universal uniform of authority, while Hutchings carried a briefcase of tools. The hospital was quiet, the dimly lit halls lending an air of mischief to their mission.

In the empty lobby, Hutchings crouched by the telephone booth, tools in hand, while Apgar stood guard. The metal scraping against wood echoed faintly as Hutchings worked, carefully prying the shelf loose. Just as they removed it, they realised the replacement shelf they'd brought was too long. Undeterred, they retreated to the women's lounge to make adjustments. Hutchings hid in a bathroom stall, using the toilet as a makeshift workbench, while Apgar stood guard outside.

"It's the only time repairmen can work in there," Apgar reassured a passing nurse who had heard the commotion, her authoritative tone masking the moment's tension.

After what felt like an eternity, their mission was complete. The liberated curly maple eventually became the back of a viola, a beautifully crafted instrument. The mission was a success—not just for the instrument it produced, but for what it represented: curiosity, rebellion, joy, and a refusal to accept limitations, no matter how mundane.

That spirit ran through everything Apgar did.

Virginia Apgar's life wasn't confined to hospital corridors or lecture halls. Her interests spilled out in every direction—vivid, eclectic, and pursued with the same unstoppable energy that defined her career.

She was an enthusiastic gardener, known for coaxing vibrant blooms from the soil and cultivating her very own orchid hybrid—the *Apgar orchid*. Her love of the outdoors ran deep. She chased salmon in Scottish rivers and explored coral reefs off the coast of Australia. She waded into rivers with the same curiosity she brought to operating rooms, casting lines with quiet focus and contagious delight.

And then there was flying.

In her 50s, she decided—quite offhandedly—to learn to pilot an aeroplane. There was no practical reason. She just wanted to see if she could. She nursed a mischievous ambition to one day fly under New York's *George Washington Bridge*. She never did. But the fact that she considered it entirely within the realm of possibility says everything you need to know.

Her quieter hobbies were no less focused. She was a devoted stamp collector, with meticulously arranged albums that charted the world one postmark at a time. Years later, when she was honoured with her own postage stamp in the *Great Americans* series, it felt like a fitting tribute—one collector among many, finally placed on the page.

She never married. She once said, half in jest, that she had simply never met a man who could keep up with her. But she wasn't solitary. Her life was threaded with connection—family, friends, collaborators. She remained close to her mother, her brother and sister-in-law, and their children. Her social circle was wide and warm, filled with fellow musicians, colleagues, and fellow mischief-makers from all corners of her world.

Her energy was legendary. She spoke quickly, her words tumbling out in overlapping waves. In his eulogy, Dr L. Stanley James captured the phenomenon with fond precision:

> *"One of the few things she could not do was talk slowly. Some people believed she had another hole for breathing. After a talk to several hundred physicians at an international meeting, it was later apparent that many had not understood a word she said—but they were enraptured and loved her. Somehow, they got the message."*

And they did. Even when the pace left listeners spinning, her meaning was unmistakable.

Her quirks added to her legend. She was known to hold two phone conversations at once—one in each hand—switching back and forth without missing a beat, a skill that left even her closest friends marvelling at her multitasking abilities. She drove, as a colleague once put it, "as if piloting a car like an aeroplane." She was notorious enough on the roads that traffic policemen attended her funeral—half in respect, half in relief, as she was their "most memorable customer."

Yet beneath the humour and high-octane presence, she was guided by something unshakeable: integrity.

She once championed an anaesthetic agent—cyclopropane—as her go-to drug for newborns. For years, she used it confidently. But when research emerged showing that it caused mild but measurable depression in infants, she didn't hesitate.

"There goes my favourite gas," she said. And that was that.

No defensiveness. No delay. Just action.

Even in the last decade of her life, Apgar showed no signs of slowing down. Alongside her public health work, she turned her attention to writing—not for scientists, but for parents. She wanted to demystify neonatal care, to make the complex clear, and to answer the questions no one else thought to ask out loud.

She co-authored a book with journalist Joan Beck titled *Is My Baby All Right?*, a practical, reassuring guide that spoke directly to families. It offered not just medical insight, but empathy—an understanding that birth, especially when things go wrong, can be terrifying. Apgar's voice in those pages was the same one she'd used at bedsides for years: brisk, honest, full of care.

Whether she was lecturing to doctors, mentoring students, or answering questions from curious teenagers, she never spoke down. She believed people could handle the truth—if it was offered clearly and with respect. That belief made her one of medicine's great communicators. She didn't obscure. She translated. She invited people in.

She died on 7th August 1974, aged 65, at *Columbia Presbyterian Medical Centre*—the very institution where she had trained, taught, led, and saved countless lives.

At her funeral, friends, family, and colleagues gathered to remember a woman who had, in every sense, lived without compromise. One of them, David Little, captured her essence with quiet eloquence:

> "The speciality of anaesthesiology lost one of its most
> distinguished ladies last year when Ginny Apgar died.
> She was a physician in every sense of the word, a true
> scientist, everybody's friend—but above all, a lady."

Apgar's legacy doesn't rest in a single innovation or a neatly summarised achievement. It lives in the delivery rooms where her score is still used daily. It lives in the safer births, in the babies who breathe because of what she saw. It lives in the generations of physicians she mentored, the systems she shaped, and the public she empowered.

She never had children of her own. But in every way that matters, Virginia Apgar was—unquestionably, unmistakably—a mother to all.

CHAPTER 6

Smile

SMILES CAN REVEAL more than you think. Let's do a quick experiment.

Grab your phone and take a picture of yourself smiling. It can be a simple, polite smile or a big, toothy grin. If you have a friend or family member nearby, ask them to join in and snap a photo of their smile.

Now, I want you to do it again—but this time, think of something genuinely funny that happened recently, or a joyful moment you shared with someone you love. If you're taking a picture of someone else, say something funny to make them laugh.

Look closely at the two smiles. Notice any differences?

In the first photo, your smile might look polite and controlled— the corners of your mouth turned up just enough to signal warmth. But in the second photo, there's something else. Your cheeks lift higher, the muscles around your eyes crinkle. There's light in the expression that wasn't there before.

This type of smile—more spontaneous, more whole—is known as the *Duchenne smile*. It's a sign of genuine happiness.

It's named after Guillaume–Benjamin Duchenne, a French neurologist who uncovered the science of true joy. Ironically, Duchenne himself didn't have much to smile about.

The Duchenne family had long been renowned for their valour and prowess at sea. Jean-Pierre-Antoine Duchenne, the stalwart commander of the *Espoir*, was a legendary figure whose exploits were immortalised in the salty tales that echoed through the docks of Boulogne-sur-Mer. Once, during a fierce battle with an English frigate, he tied himself to the ship's wheel, determined to navigate through the raging storm. With cannon fire all around him and nature unleashing its fury, he stood firm, refusing to back down. This act of defiance, among others, earned him the *Légion d'Honneur* from Napoleon Bonaparte himself in 1804. For the people of Boulogne, Jean-Pierre-Antoine wasn't just a man—he was a legend.

In stark contrast to his father, young Guillaume Benjamin Duchenne was a reserved and introspective soul. Born in September 1806 in the coastal town of Boulogne-sur-Mer, he grew up amidst the salty breeze and stories of seafaring glory. There, he would gaze at the horizon, listening to the crash of waves against rocks, and feel a strange pull— not towards the sea, but inward, to the mysteries of life itself.

While his father urged him to embrace the family's maritime legacy, Duchenne found the sea's unpredictability unsettling. The ocean was a realm of chaos, and Duchenne's soul yearned for understanding, for order—a yearning that grew as he watched his father command the seas with unflinching courage. The younger Duchenne admired his father, but he knew that his courage would take a different form.

When Duchenne turned 19, he made his decision. The salty air of Boulogne was exchanged for the structured halls of *Bishop Haffreingue's* college in Douai, where he earned his *Baccalauréat ès lettres*.

In 1827, Duchenne moved to Paris to study medicine. The city pulsed with ambition and intellect. He was mentored by some of the most prestigious figures of the time—René Laennec, the inventor of the stethoscope, and Baron Guillaume Dupuytren, whose surgical skills were widely celebrated. Duchenne's quiet demeanour set him apart as his peers competed for recognition and status. He spent long nights in the dissecting rooms, his hands steady and his mind sharp, unravelling the secrets of the human body. In 1831, he defended his thesis on burns—a respectable topic, though it did little to distinguish him in the fiercely competitive world of Parisian medicine.

Shortly after his graduation, tragedy struck.

A letter arrived from Boulogne-sur-Mer, the words smudged as if the writer's tears had mingled with the ink. Duchenne's hands trembled as he broke the seal. As he read the news of his father's passing, a sharp ache shot through him, leaving him breathless. The room seemed to spin, and he clutched the back of a chair to steady himself. For a moment, he stood frozen, staring at the words that had shattered his world.

The loss felt like a shipwreck in his soul; the foundation of his identity was gone in an instant.

Duchenne returned home to find his family adrift in grief. He took on the role of a family doctor in his hometown. In Boulogne-sur-Mer, he found his place—not among the waves, but in the quiet resilience of everyday life, helping those around him with the same spirit that had once guided his father through stormy seas.

For a time, joy returned.

In December 1831, Guillaume Duchenne married Barbe Boutroy, a woman whose kindness and quiet strength mirrored his own. Their union brought a sense of stability, and when Barbe became pregnant, Duchenne's heart swelled with anticipation. He imagined their child's future, the life they would build together. As a family doctor, he had delivered countless babies, and the thought of welcoming his own child into the world brought him immense joy.

When the time came in January 1833, Duchenne was there, guiding the delivery of his son, Guillaume Maxime. It was a moment of profound happiness and accomplishment.

But fate had other plans.

Two weeks after the birth of their son, Barbe died from puerperal sepsis (a severe infection following childbirth).

Duchenne's world shattered.

In the weeks that followed, grief consumed him. He blamed himself for Barbe's death, tormented by the thought that his role in her delivery had somehow contributed to her suffering. In his vulnerable state, he allowed his mother-in-law to care for the baby. She, grieving and furious, held him responsible. She forbade any contact with his son.

He would not see Guillaume Maxime again for thirty years.

This tragedy caused a rift between Duchenne and the community. The whispers and mistrust led him to step back from his practice. He sought solace in his home, burying himself in books and finding comfort in the music of Bach and Beethoven, which he played on his violin. The violin became an extension of his grief—its mournful melodies echoing the ache he could not put into words.

As years passed, age turned Duchenne bald, leaving him with thick sideburns that framed his thoughtful face. Gradually, he began to return to work, his spirit slowly rebuilding.

In 1839, he remarried. Honorine Lardé, a vivacious young widow, brought light back into his life. Her extroverted personality stood in contrast to Duchenne's quiet, reclusive nature, yet that difference sparked something between them. She was warm, social, outward-looking. Where he withdrew, she reached out. Her energy didn't diminish his—it drew him gently back into the world.

Around the same time, Duchenne's curiosity stirred once more. He began exploring *electropuncture*, a procedure that combined the ancient principles of acupuncture with the emerging science of electricity. It wasn't a fashionable area of study, and few in Paris considered it anything more than medical eccentricity. But for Duchenne, it opened a door.

He began experimenting with *electropuncture* in 1835. His first patient was a fisherman plagued with neuralgia—nerve pain that flared unpredictably and left the man unable to work. Duchenne applied a mild electric current to the man's face and watched as the affected muscles flickered and twitched in response.

That moment—the visible contraction, the involuntary motion, the sudden quieting of pain—sparked something in him.

Electricity wasn't just force. It was communication. It moved through the body like language, revealing hidden dysfunctions and untapped pathways.

Duchenne's inventive mind quickly took over. He designed a portable machine housed in a mahogany case, equipped with an induction coil and humid surface electrodes. These devices allowed him to stimulate individual muscles with far greater precision, avoiding the pain and damage often caused by needles. It was a leap forward—cleaner, safer, more controllable.

He called the technique *localised faradisation.*

For the first time in years, Duchenne felt a sense of forward motion—a purpose that might rise above his personal losses. His grief was not gone, but he could carry it alongside something meaningful.

In 1842, with his device refined and his resolve sharpened, Duchenne packed his belongings and set off once again for Paris.

When Guillaume Duchenne arrived in Paris, he was alone and without funds—a provincial doctor armed only with an unusual device and a belief in its potential. To distinguish himself from the many other physicians named Duchenne, he added the suffix *"de Boulogne".*

But even that didn't open doors.

The city, with its grand hospitals and glittering reputations, could be as cold as its winters. Duchenne moved from one hospital to the next, offering his methods for treating neuromuscular diseases. The receptions were polite at best, often patronising, and more often still, dismissive. The Parisian medical elite saw him as little more than an eccentric from the provinces.

One day, while demonstrating his device at a clinic, he overheard a surgeon murmur, "Another country quack, chasing his delusions." The words landed hard. But Duchenne, ever inward, reminded himself of his father—tied to the ship's wheel in the storm, refusing to surrender. He would do the same.

His invention—a portable machine composed of a battery and an induction coil, neatly stored in a mahogany case—was unlike anything most physicians had seen. With it, Duchenne could stimulate individual muscles with pinpoint accuracy.

It wasn't just an experiment. It was a method—a tool for decoding the hidden messages of neuromuscular diseases (conditions that affect the muscles and the nerves controlling them, leading to issues like weakness, spasms, or paralysis).

He began documenting what he saw: when a muscle responded to stimulation, it suggested a problem within the central nervous system—the brain or spinal cord. When it didn't, the issue was likely in the muscle or nerve itself. These insights were quiet revelations. But at the time, they fell on deaf ears.

Still, Duchenne remained undeterred.

He cut a distinctive figure in the hospital wards of Paris. Medium height, solid build, and with slow, deliberate speech softened by his provincial accent, he carried both humility and an unshakeable focus. Patients and staff alike noticed his habit of rising early, moving between hospitals, always seeking out the most puzzling cases. He was tireless. And though many scoffed, few could ignore him for long.

Then, during one of those visits, a young intern named Jean-Martin Charcot took notice.

Intrigued by Duchenne's strange device and methodical approach, Charcot invited him to work at the *Salpêtrière*—a sprawling hospital complex then known more for housing the chronically ill than for cutting-edge research. Duchenne was assigned to its quieter wards, places others avoided.

But he found something there that suited him perfectly: time, space, and freedom.

In those forgotten corners, surrounded by patients dismissed by others, Duchenne began to refine his techniques. He gathered clinical evidence. He took notes obsessively. He treated those no one else would touch. And without fanfare, he laid the foundation for something far greater.

Charcot, inspired by Duchenne's work, would later go on to become one of the most important figures in medical history—the *Father of Neurology*. But in those early days, he was still learning. And Duchenne, with his quiet dedication and radical methods, was his teacher.

From the patients' perspective, Duchenne was an odd but oddly comforting presence. He'd appear in the corridors, carrying his mahogany case like some mysterious travelling salesman. The women of the *Salpêtrière*, long forgotten by the system, began to whisper:

"Here comes the little old man with his mischief box."

There was amusement in their tone—but also hope.

They had suffered for years, often in silence. Duchenne offered something new. Relief. Curiosity. Attention. A sense that they were, once again, seen.

As Duchenne applied his treatments—flickers of electricity coaxing contracted muscles into temporary ease—he began to notice something else.

A pattern.

The smiles.

They appeared at moments of relief, yes—but there was more to them. Certain smiles, Duchenne noticed, involved the eyes as well as the mouth. They weren't just physical reactions. They were emotional signatures. Authentic joy revealed itself in the body.

He couldn't stop thinking about it.

What if, he wondered, the muscles of the face held the key to understanding human emotion?

The idea gripped him. Quietly, completely.

Duchenne's fascination with human emotions led him to test various facial expressions using his innovative electrical stimulation techniques. He observed that genuine emotions elicited specific muscular responses, particularly in the face. Duchenne explained that the zygomaticus major, a muscle responsible for lifting the corners of the mouth, and the orbicularis oculi, which crinkles the skin around the eyes, worked together to create an authentic smile. He wrote:

> "The emotion of frank joy is expressed on the face by the combined contraction of the zygomaticus major and the inferior part of the orbicularis oculi. The first muscle obeys the will, but the second... is only put at play by the sweet emotion of the soul. Finally, fake joy, the deceitful laugh, cannot provoke the contraction of this latter muscle."

This distinction became known as the *Duchenne smile*—an authentic expression of joy.

As his work progressed, Duchenne became increasingly captivated by the emerging field of photography. He realised that sketches could not fully capture the precision and subtlety of muscle movements and began to see photography as essential to his research. In 1862, he reflected:

"From 1852 onwards, I had the idea of illustrating, with the help of this wonderful procedure [photography], the specific action of individual muscles through electrical faradisation... This convinced me to learn and study the art of photography from the point of view of its application to physiology and pathology."

The sharp clarity of photographs allowed him to document the effects of electrical stimulation with unprecedented accuracy, creating a lasting visual record of his work.

Despite his enthusiasm, Duchenne faced challenges. Many patients found facial stimulation unpleasant, often experiencing painful spasms. This limitation frustrated him, as it constrained his ability to fully explore the nuances of expression.

A breakthrough came, however, when he met an old cobbler at the *Salpêtrière*. The man, with a toothless, thin face and otherwise unremarkable features, suffered from near-total facial anaesthesia (loss of sensation in the face). Duchenne saw this as an extraordinary opportunity.

He approached the man, who sat quietly in a corner, absorbed in his own thoughts.

"Good day," Duchenne greeted him, his voice a blend of warmth and curiosity. "I have a proposition for you, my friend. I understand you cannot feel much sensation in your face?"

The cobbler looked up, his eyes weary but attentive. "Aye, Doctor. It has been like this for years. Feels as though I've a mask on, always."

Duchenne nodded thoughtfully. "I've been working on a study involving electrical stimulation of the facial muscles. Many find it painful—but in your case, you might feel nothing at all. Would you be willing to assist me with my experiments?"

The cobbler chuckled, revealing a toothless grin. "Well, Doctor, it helps others and you promise it won't hurt, why not? I've nothing to lose."

Duchenne smiled in return, his eyes lighting up. "Thank you. Your assistance could lead to something remarkable."

The cobbler's unique condition allowed Duchenne to stimulate individual facial muscles without causing discomfort. Over the following weeks, he documented a wide range of expressions—from joy to anguish—with unparalleled precision. The cobbler, who at first viewed the experiments with mild amusement, began to take pride in his role.

"If I can lend my old face to science, why not?" he quipped one day, earning a hearty laugh from Duchenne.

These experiments culminated in the publication of Duchenne's now-infamous photo album, *Album de photographies pathologiques*, in 1862. The album—the first medical treatise illustrated with photographs of living patients—showcased his findings in vivid detail.

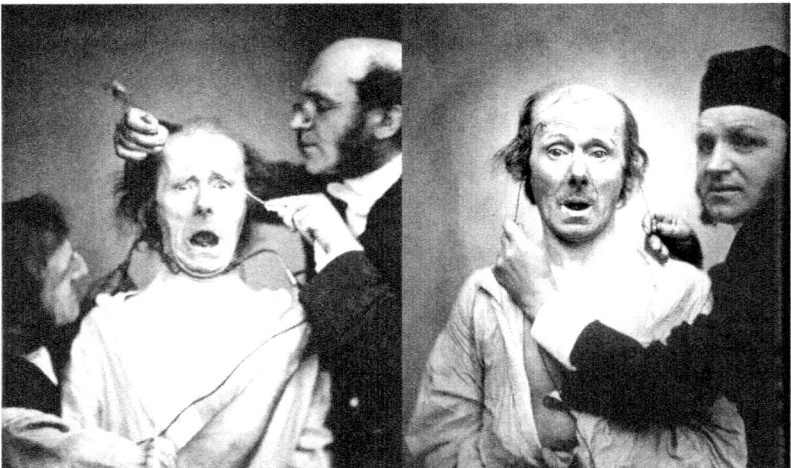

On the left, Dr Duchenne de Boulogne and his assistant electrically stimulate the facial muscles of an elderly cobbler afflicted with total facial anaesthesia. The cobbler's exaggerated expressions highlight the specific muscles involved in different emotional states. On the right, Dr Duchenne applies electrical stimulation to the forehead muscles of the same cobbler, who remains pain-free.

This collection of early photographs by Dr Duchenne de Boulogne demonstrates the mechanics of facial expressions. The upper and lower rows depict patients with different expressions on either side of their faces. These images illustrate the effect of electrical stimulation on specific facial muscles, providing insight into the diverse range of human emotions and their underlying physiological mechanisms.

Each image captured the relationship of muscles and emotions, creating a visual language that bridged science and art. Duchenne's work did not go unnoticed. Among those who admired his findings was Charles Darwin, who praised Duchenne's meticulous research on human physiognomy. Darwin incorporated several of Duchenne's photographs into his work, *The Expression of the Emotions in Man and Animals*, published in 1872. This collaboration underscored

Duchenne's influence and cemented his legacy in both medicine and evolutionary science.

As Duchenne's reputation grew, young doctors flocked to learn from him. His influence in Paris expanded, and his friendship with Jean-Martin Charcot deepened as they pushed the boundaries of neurological research together.

As Duchenne carefully prepared his equipment for another demonstration one afternoon, a young doctor entered the room. He hesitated by the doorway, his hands clasped tightly in front of him, before finally stepping forward.

"Excuse me, Dr Duchenne," the man began, his voice trembling slightly. "My name is Guillaume Maxime." He paused; his eyes fixed on Duchenne's face. "I believe... I am your son."

Guillaume Maxime had not been allowed any contact with his father since he was two weeks old, remaining under the strict watch of his maternal grandmother. After she died in 1851, he experienced a sense of freedom he had never known. Moving from Boulogne to Strasbourg, he pursued a career as an army physician, which eventually took him to distant lands. For many years, he served as a senior military doctor in North Africa and Turkey, finding purpose in his work while remaining unaware of his father's groundbreaking contributions to medicine back in France.

Meanwhile, Duchenne's curiosity drove him to numerous discoveries. Among his most notable was identifying a condition he called *pseudohypertrophic muscular dystrophy*, now recognised as *Duchenne muscular dystrophy*. One day at the *Salpêtrière*, Duchenne noticed a young boy struggling to walk. The boy's movements were laboured, his steps heavy and uncoordinated. Duchenne's sharp eye was drawn to the boy's calves, which appeared unusually large.

Observing the child with sympathy, Duchenne wondered why these swollen muscles failed to provide strength.

Intrigued, he decided to investigate further. Gently examining the boy's legs, he noted the firm, swollen appearance of the calves. He hypothesised that the enlargement wasn't due to actual muscle growth, but rather something abnormal. Eager to uncover the truth, he developed an innovative tool—a needle system he called *Duchenne's harpoon*. This allowed him to take small muscle tissue samples without anaesthesia, a pioneering precursor to modern biopsy techniques.

When Duchenne analysed the samples, he made a startling discovery: the apparent muscle growth resulted not from healthy tissue, but from an accumulation of fatty and connective tissue replacing muscle fibres. This explained the paradox of the boy's large calves and significant weakness.

Duchenne's findings helped distinguish this condition from other muscular diseases, providing a more precise understanding and laying the foundation for future diagnostic methods.

His work didn't stop there. His passion for studying neuromuscular diseases led him to identify and describe other conditions. For example, he documented progressive spinal muscular atrophy, where muscles waste away over time, leading to weakness and paralysis. He also detailed observations of progressive locomotor ataxia, a disorder that impairs movement due to nerve damage.

One early morning, Duchenne stood at his window, peering out at the quiet Parisian street below. The faint sound of uneven footsteps on the cobblestones caught his attention. Pressing closer to the glass, he squinted at a group of patients walking toward Lamalou's baths. Their distinctive gait made his heart race.

"Ataxic patients!" he exclaimed, his breath fogging the windowpane.

Excitement surged as he recognised their exaggerated leg movements. Without hesitation, Duchenne rushed from his room, bursting into the apartment where he was staying.

"Everyone, wake up!" he shouted, his voice ringing through the halls. "You must see this!"

Groggy and bewildered, the room's occupants—including his host—stirred and looked at him in confusion.

"What is it, Duchenne? It's barely dawn."

"There are patients outside, walking with that distinctive gait," Duchenne explained, mimicking the movements by throwing his legs forward in exaggerated steps. "They belong to me. They are ataxic patients—and I must question them all!"

With that, he dashed into the street. His host, shaking his head with a mixture of amusement and exasperation, muttered, "Only Duchenne would get this excited before breakfast."

Approaching the patients, Duchenne introduced himself, his enthusiasm disarming.

"Good morning! I am Dr Duchenne de Boulogne. May I speak with you about your walking difficulties?"

The patients, intrigued by his energy and genuine interest, agreed, providing him with invaluable insights for his research.

Duchenne's discoveries had far-reaching impacts. His technique for phrenic nerve stimulation paved the way for artificial respiration. His methods of electrical stimulation to relieve pain evolved into today's transcutaneous electrical nerve stimulation (TENS). His early experiments regulating heart rhythms through electric shocks would also become precursors to modern cardiac pacemakers.

In 1862, just when Duchenne thought life couldn't get any better, his world took a remarkable turn. Guillaume Maxime, having settled in Paris and chosen to follow in his father's footsteps as a neurologist, sought him out. For years, he had admired his father's work from afar. His own medical career had deepened his understanding of Duchenne's significance.

Their reunion was emotional.

When Guillaume Maxime first approached him, he hesitated at the doorway, unsure how his father would react. Taking a deep breath, he stepped forward.

"Excuse me, Dr Duchenne," he began, his voice trembling slightly. "My name is Guillaume Maxime." He paused, meeting his father's gaze. "I believe... I am your son."

Duchenne froze, his hands still resting on his equipment. The words hung in the air, their weight reverberating through decades of separation and loss. Slowly, he turned, his voice barely a whisper.

"Guillaume Maxime... my son?"

Tears welled in both men's eyes as the years of distance melted away.

In the weeks that followed, the bond between father and son grew strong. Duchenne eagerly shared his research and techniques, his pride evident in every interaction. Guillaume Maxime, in turn, soaked in his father's knowledge, inspired by his brilliance.

Their collaboration culminated in Guillaume Maxime defending his thesis on atrophic paralysis of childhood, a condition closely related to Duchenne's research. Standing before a panel of esteemed physicians, he concluded his defence by dedicating his work to his father. Watching from the audience, Duchenne felt an overwhelming sense of pride.

As the applause filled the room, Duchenne's face broke into a genuine, heartfelt smile.

A Duchenne smile.

The French–Prussian War of 1870–71, a brutal conflict between France and the German states led by Prussia, marked a devastating period in French history. It culminated in the fall of Napoleon III, the siege of Paris, and the eventual unification of Germany. For Duchenne, it marked a personal tragedy at what had seemed the height of his happiness.

As the Prussian army prepared to besiege Paris, his wife Honorine fled to England for safety. Duchenne, ever the physician, chose to stay behind—committed to caring for wounded Parisians despite the looming danger.

Tragedy struck in December 1870. Duchenne received the devastating news that Honorine had died in England. Still reeling from her loss, he was shattered once more less than a month later when his beloved son, Guillaume Maxime, succumbed to typhoid fever.

The back-to-back blows silenced something in him.

Overwhelmed by grief, Duchenne withdrew from his work. The journals that had once been his anchor sat untouched on his desk. His instruments, so carefully maintained, gathered dust. Though his name still carried weight across Europe, the man himself seemed lost.

He travelled—visiting Austria, England, Spain. Colleagues welcomed him warmly, and he was honoured by foreign medical societies. But none of it brought comfort. Those closest to him remarked that his once-curious gaze had dimmed, and some whispered that he would never smile again.

In his solitude, Duchenne did something unusual. He began hosting dinner gatherings—quiet affairs, often understated, where acquaintances, colleagues, and former students were surprised to find themselves invited by the once-reserved doctor. He offered wine, shared anecdotes, and tried to stir a sense of conviviality. But those who sat at his table noticed how his efforts to smile were fleeting. The gestures were there. The light in his eyes was not.

The evenings were a distraction at best. A brief reprieve from the absence that followed him from room to room.

In August 1875, Duchenne suffered a stroke—a final, cruel twist. His dear friend Jean-Martin Charcot rushed to his side, not just as a physician but as a companion. He stayed with him, sleeping in a chair beside his bed for several nights, refusing to let him be alone.

Despite his failing health, Duchenne remained lucid. Even as his body weakened, he continued to scribble notes in the margins of his old journals. His hand trembled, but his mind still reached for the same questions.

On 15th September 1875, just two days before his 69th birthday, Duchenne de Boulogne died.

Duchenne's life, marked by profound scientific contributions and personal tragedy, came to a sombre close.

What is Duchenne muscular dystrophy?

Duchenne muscular dystrophy (DMD) is a genetic disorder that weakens muscles and causes them to break down over time. It primarily affects boys and often becomes noticeable in early child-hood. Children with DMD may have difficulty walking, running, or climbing stairs, and as they grow older, they may need braces or a wheelchair to move around. Over time, the disease can also weaken the muscles needed for breathing and heart function, making it a serious and life-threatening condition.

DMD is caused by a mutation in a gene responsible for producing dystrophin, a protein that helps keep muscles strong and healthy. Without enough dystrophin, muscle fibres are gradually replaced by fat and connective tissue, leading to the loss of strength and func-tion. While there is currently no cure for DMD, advances in medical research are helping doctors manage symptoms and improve quality of life.

CHAPTER 7

A New Sound

ON A CHILLY afternoon in September 1816, René Laennec wandered the parks of Paris, grappling with a gnawing frustration. Earlier that morning, he had examined a young woman suffering from breathlessness, irregular heartbeats, and debilitating fatigue—symptoms suggestive of a cardiac condition. Yet diagnosing her had proven nearly impossible.

The standard method of auscultation—listening to the sounds of the heart and lungs by pressing one's ear to the patient's chest—had served physicians since ancient Greece. Hippocrates himself had relied on it. But by the early 19th century, it was increasingly clear how flawed this approach was. For some patients, particularly those who were overweight, chest sounds were faint and muffled, almost indiscernible. Hygiene posed another challenge; physicians frequently encountered unwashed bodies and, disturbingly, vermin. These discomforts made the practice unsavoury at best, impractical at worst—especially in the crowded hospitals of Paris.

In this young woman's case, her considerable size rendered the method useless. Laennec found himself unable to hear her heart's faint murmurs through layers of tissue. Modesty further complicated matters. For a respectable young woman, the thought of a doctor pressing his ear to her chest—even in a clinical setting—was an awkward intrusion. The barriers were physical, cultural, and inherently personal. Laennec had left the encounter not with answers, but with a growing sense of futility.

To clear his mind, he took a walk through the gardens of the *Louvre Palace*. The early signs of autumn coloured the scene: leaves turning gold and amber, a crisp breeze carrying the scent of damp earth. Yet Laennec noticed none of it. The distant clatter of carriages and the chatter of Parisians out strolling only heightened his irritation.

He was so absorbed in his thoughts that he almost missed the burst of laughter from two boys playing nearby. One held a long wooden stick and said, "Watch this! Place your ear at the other end." The second boy obliged, pressing his ear to the stick. Then the first scratched one end with a pin. The listening boy's face lit up in wonder.

"I can hear it!" he exclaimed, amazement clear in his voice.

Laennec stopped in his tracks, curiosity piqued. He watched as the children repeated the trick, and the idea struck him: the sound travelled through the stick—amplified. As a skilled flautist who often crafted his own instruments, Laennec was no stranger to the principles of acoustics. He understood how vibrations moved through solid materials. And now, standing in the gardens, watching a child's simple experiment, the pieces came together.

What if this principle could be applied to auscultation? A hollow tube—a simple device—might bridge the physical gap between doctor and patient, amplifying heart and lung sounds without distortion. He could almost see it: a tool that made diagnosing patients like the young woman possible, one that would circumvent the limitations of immediate auscultation.

The frustration that had dogged him all day dissolved in a rush of inspiration. Laennec turned sharply on his heel, heading back towards the hospital with a renewed sense of purpose. The gardens, with their golden leaves and carefree children, had offered more than a momentary escape; they had sparked a revolution in medical practice. By day's end, he would begin work on what would become one of medicine's most transformative inventions: the stethoscope.

René Laennec was born in 1781 in Quimper, Brittany, a town known for its winding cobblestone streets and the gentle flow of the Odet and Steir rivers. Though picturesque, the charm of his hometown stood in contrast to the harsh realities that shaped his childhood.

When René was just five, his mother succumbed to tuberculosis—a disease that haunted so many families of the era. The loss left a void in the boy's life that his father, a distant and preoccupied lawyer, seemed unable—or perhaps unwilling—to fill. Laennec was devastated by his mother's passing, though he could not yet grasp the nature of tuberculosis. He simply knew she was gone.

Unable to care for him, his father sent Laennec to live with his great-uncle, Abbé Laennec, a priest. The abbey offered a world far removed from the bustle of Quimper. Its quiet halls, scented with candle wax and ageing wood, provided a peculiar sort of solace. Here, Laennec began to grow, not just in body, but in spirit. Though often unwell—prone to recurring fevers and fatigue—he found joy in small, grounding moments: the sweet, melodic notes of the flute he learned to play, the soft curl of wood shavings under his knife as he practised carving. In music and craftsmanship, Laennec discovered a way to transcend his frailty.

His studies with Abbé Laennec deepened his inquisitiveness. Latin and Greek texts became portals to worlds of knowledge and imagination. By 12, his mind was sharp, but once again, life pulled him from the familiar. He was sent to Nantes to live with another uncle—Dr Guillaume François Laennec, the Dean of the Faculty of Medicine at the university.

Nantes was alive with revolutionary fervour. The ideals of liberty and equality swept through its streets, accompanied by bloodshed and upheaval. For the young Laennec, this new world was both thrilling and terrifying.

Under his uncle's watchful eye, Laennec began to flourish. Stern but kind, Dr Laennec became the father figure René had never truly known. The *Hôtel-Dieu*, the city's hospital, soon became a second home. Its corridors echoed with the moans of the sick and wounded, its air heavy with the metallic scent of blood and illness. Here, in the raw theatre of suffering, Laennec found his calling.

At just 14, he assisted in patient care, his hands trembling as he dressed wounds or prepared instruments for bloodletting. This was no classroom—it was life, exposed and unflinching.

By 18, he had earned a position as a third-class surgeon at the military hospital. Though still young, he held steady amid the cries of soldiers and the chaos of war. But just as his confidence began to grow, he encountered a familiar obstacle: his father.

Now remarried, Laennec's father disapproved of his son's medical aspirations and demanded he abandon his studies. The blow was devastating. Laennec left Nantes, retreating to the countryside, despondent. He sought solace in long walks, Greek poetry, and the melodies of his flute—but the ache for medicine never left him.

Recognising this unshakable passion, Dr Guillaume intervened. "You have the hands of a surgeon and the mind of a physician," he told him. With his uncle's support, René defied his father and set his sights on Paris—the epicentre of medical progress.

In 1800, Laennec arrived in the capital. Paris was electric with revolutionary energy and scientific ambition. He joined the *École Pratique*, Europe's foremost centre for clinical research. There, he encountered the formidable Guillaume Dupuytren—a surgeon known as much for his brilliance as his tyranny.

Dupuytren's obsession with anatomical precision demanded rigour from his students. Laennec, driven by his own perfectionism, thrived under the pressure. Within a year, he had won prizes in both medicine and surgery. His aptitude stood out, even in the cut-throat world of Parisian medicine.

It was during this period that Laennec made a pivotal observation. Among his patients, he noted dark, spreading tumours that he suspected were more than just deposits from tuberculosis or coal dust. Drawing on his classical training, he named them *"melanose"* from the Greek word for "black."

In 1804, while still a medical student, he delivered the first recorded lecture on melanoma. He explained that these dark lesions represented a metastatic skin cancer, not just superficial discolouration. The lecture, published in 1806, marked a significant advancement in cancer research.

But the recognition was not without controversy. Dupuytren, never one to share credit easily, felt his own contributions had been eclipsed. The rift between mentor and student deepened, leaving a bitter taste that would linger.

Laennec's gift for naming did not end there. He also described cirrhosis, a liver disease marked by tawny, yellow nodules. Though the disease itself was already recognised, Laennec coined the term "cirrhosis" from the Greek *kirrhos*, meaning yellowish. In time, it would become eponymously linked to him: *Laennec's cirrhosis.*

Laennec studied under Jean-Nicolas Corvisart, Napoleon's personal physician and a towering figure in clinical medicine. Corvisart's lectures on bedside diagnosis captivated him—particularly his revival of a technique long overlooked: percussion.

The story of percussion's origins fascinated Laennec as much as its practical use. In the 18th century, a Viennese physician named Leopold Auenbrugger had pioneered the method by observing his father tap wine barrels to estimate how full they were. Like a barrel, he reasoned, the human chest could produce different sounds depending on what lay beneath the surface. A hollow note suggested healthy, air-filled lungs; a dull sound might indicate fluid, masses, or other obstructions.

Auenbrugger documented his findings in *Inventum Novum*, a slim volume that went largely unnoticed in his lifetime. Yet there was genius in its pages. An accomplished musician, Auenbrugger likely had an ear trained to detect nuance—a gift that served him well in diagnosing pulmonary and cardiac abnormalities through variations in sound. His method could detect dullness over the heart in cases of pericardial effusion (a buildup of fluid in the sac surrounding the heart),or signal cardiac enlargement when percussion notes changed.

But Auenbrugger's brilliance went unrecognised. His work languished in obscurity until Corvisart, recognising the value of this technique nearly 50 years later, translated *Inventum Novum* into French and brought percussion back into the medical mainstream. His passion for diagnostic innovation was infectious, and it lit a fire in Laennec. He watched, fascinated, as sound revealed the body's hidden truths. For Laennec, percussion was more than a technique—it was a revelation. It proved that sound, interpreted correctly, could become a language of the sick body.

This insight stayed with him, slowly shaping his view of medicine. It was not lost on him that Auenbrugger, like himself, had a musical background. There was something more than coincidence in that. To Laennec, music and medicine shared a principle: the disciplined listening for patterns, for anomalies, for meaning embedded in resonance.

At the intersection of sound and diagnosis, the seeds of Laennec's greatest invention were quietly taking root.

🩺

Tuberculosis loomed over René Laennec's life, an ever-present shadow that shaped his personal and professional journey. It had taken his mother when he was just a boy, a loss that left an unhealed wound. Years later, the disease would claim others he held dear; his brother and his beloved uncle, Dr Guillaume Laennec, the man who had been a father to him and the driving force behind his medical career. The grief from these losses was overwhelming, and though Laennec rarely spoke of it, those close to him could see the weight he carried. He was no stranger to the disease himself, battling tuberculosis even as he worked tirelessly to understand it.

During his medical training, Laennec formed a close bond with Gaspard Laurent Bayle, a respected tuberculosis specialist whose intellect and dedication matched Laennec's own. Their shared fascination with the disease drew them together. Late at night, over flickering lamplight, they would discuss their findings and ideas. These conversations became a source of inspiration for Laennec, reinforcing his motivation to uncover the mysteries of tuberculosis.

Even after graduating, the disease continued to dominate his thoughts. Despite his frail health, exacerbated by asthma and his exposure to tuberculosis patients, he threw himself into clinical practice, teaching, writing, and editing medical journals. His frame—small, lean, and constantly worn—belied the formidable mind within. His students and colleagues recognised it: behind the reserved exterior was a brilliant, driven physician whose dedication bordered on self-sacrifice.

But sorrow followed him with unnerving persistence. When his uncle died of tuberculosis, the loss nearly shattered him. Guillaume had been his foundation, the one who had recognised his talent and nurtured it. Losing him felt like losing part of himself. The death of his brother soon after deepened the wound. Drained emotionally and physically, Laennec withdrew from Paris and sought refuge in the quiet hills of Brittany.

There, he found temporary solace. The landscape of his youth, with its rolling pastures and sea-salted breezes, offered a kind of gentle anonymity. He took long walks, played his flute for hours, and let the rhythms of the natural world soothe his spirit. But even in retreat, he could not escape the persistent question: *Why tuberculosis?* Why had it taken so many? And why, despite all his knowledge and devotion, had he been powerless to stop it?

His time in Brittany gave him space to grieve—and, slowly, to heal. With the passing of months, his strength returned. So too did his resolve. He would not let loss define him. There was still work to be done.

When he returned to Paris, Laennec plunged back into his career. He accepted an editorial role at the *Journal de Médecine*, expanded his private practice, and intensified his research on chest diseases. But despite his growing reputation, he struggled to secure a senior physician's post at any of the city's major hospitals. The rejection stung, but he pressed on.

Then came the war years—1812 and 1813—bringing with them a different kind of chaos. The city was a whirlwind of activity, filled with the sounds of marching soldiers, clattering carriages, and the ever-present tension of conflict.

These years were marked by Napoleon's disastrous Russian campaign in 1812 and the subsequent battles throughout 1813, culminating in the critical *Battle of Leipzig*, where Napoleon's forces were decisively defeated by the Sixth Coalition (a military alliance of Austria, Prussia, Russia, Sweden, the United Kingdom, and other nations).

Hospitals were overwhelmed with wounded soldiers, and resources were stretched thin. At the *Salpêtrière* Hospital, Laennec took charge of the wards, tending to Breton soldiers from his home region. The work was gruelling, and the days were a blur of chaos and fatigue. The cries of the injured and the clatter of carts echoed through the corridors.

Still, amid the clamour, Laennec's thoughts often drifted back to tuberculosis. Even as he dressed wounds and managed emergency care, the quiet moments—the pause at a soldier's bedside, the late-night reflections—were filled with the same haunting questions. What *was* this disease, truly? And how could it be stopped?

When Napoleon abdicated in 1814 and peace briefly returned to France, Laennec refocused. But heartbreak struck again. In 1816, his friend and colleague Bayle died—yet another victim of the disease they had worked so hard to understand.

The loss devastated Laennec, but it also shifted his focus. He would not let tuberculosis continue unchecked. Not if there was something more he could do.

Shortly after Bayle's death, Laennec was appointed as a physician at *Necker Hospital*. The post gave him what he had long sought: time, space, and institutional support to dedicate himself fully to the study of chest disease. He spent countless hours at the bedside and in the dissection room. His autopsies revealed a startling truth—tubercle lesions weren't limited to the lungs. They could be found in nearly every organ. This insight expanded the understanding of tuberculosis significantly.

And yet, even with all he had learned, Laennec still failed to recognise the disease's infectious nature. Like so many of his time, he viewed it as constitutional, not contagious. The mystery, in part, remained unsolved.

But one problem continued to haunt him above all: the physical difficulty of examining the chest. Despite his encyclopaedic knowledge, despite his detailed observation, he was still constrained by the limitations of traditional auscultation.

Then came September 1816. The case of the overweight young woman, the faint murmur he could not hear, the gnawing frustration—and the walk through the gardens. The laughter of two boys. A wooden stick.

And with it, inspiration.

Upon returning from his walk, inspired by the boys' game with the stick, Laennec acted immediately. He tightly rolled a sheaf of paper into a cylinder, placed one end on the woman's chest, and pressed his ear to the other.

What he heard astonished him: the faint murmurs of her heart, clear and distinct—more audible than they had ever been through direct contact.

Reflecting later, he wrote:

> *"Not a little surprised and pleased to find that I could thereby perceive the action of the heart in a manner much more clear and distinct than I had ever been able to do by the immediate application of my ear."*

Thrilled by the possibilities, Laennec began refining the idea with the same resourcefulness that had marked his childhood. His early experiences with woodcarving and musical instruments proved invaluable as he experimented with different materials: paper, wood, even Indian cane. He found that a narrow central channel significantly improved sound transmission.

After much trial and error, he arrived at a design—simple yet effective: a wooden cylinder, 25 centimetres long and 3.5 centimetres wide, with a funnel-shaped cavity to enhance amplification. The device could be separated into two parts for easier transport.

Laennec named it the *stethoscope*, from the Greek *stethos* (chest) and *skopein* (to examine). It was more than a tool—it was the opening note of a new medical language.

With the invention came a new vocabulary. Laennec's command of languages gave him the precision to define not just an instrument, but an entire methodology. From the Latin *auscultare*, meaning "to listen attentively," he coined the term *auscultation* to describe the process of using sound in diagnosis.

His observations using the stethoscope enabled him to accurately identify and classify different breath sounds. He introduced terms like *rales* (a rattling sound in the lungs), *rhonchi* (a low-pitched, snoring-like sound), *crepitance* (a crackling sound often associated with fluid in the lungs), and *egophony* (a peculiar change in voice sounds). These terms are still used today.

These were not abstractions—they were acoustic signatures of disease. And Laennec, in correlating them with his meticulous autopsy findings, began linking sound to pathology with unprecedented accuracy.

He identified and described, with clinical precision, a host of respiratory and cardiac diseases: tuberculosis, pneumonia, bronchitis, pleurisy, emphysema, pneumothorax, pulmonary oedema. No longer would these conditions be guessed at by surface signs alone; now, they could be *heard*, mapped, and understood.

In 1818, Laennec presented his invention and findings to the *French Academy of Sciences*. At first, his peers reacted with scepticism. Many senior physicians were hesitant to embrace the new device, preferring the methods they had long trusted. Some dismissed it outright as a novelty.

But Laennec was undeterred. He demonstrated the stethoscope's utility repeatedly, coupling careful auscultation with clinical and post-mortem findings. His clarity, his persistence—and the undeniable results—began to turn heads.

He began selling the instrument alongside his written work, and gradually, adoption followed. Younger doctors, in particular, were eager to explore this promising innovation.

The stethoscope's reach soon extended beyond France. One of its early champions abroad was Thomas Hodgkin, a brilliant young physician at *Guy's Hospital* in London. Impressed by the instrument's diagnostic potential, Hodgkin brought it back from Paris in 1822.

The reaction from his older colleagues was cool, even mocking. At one point, the stethoscope was reportedly repurposed as a flowerpot in the hospital's common room—its hollow cylinder filled with blooms and left standing upright on a table.

But Hodgkin remained undaunted. When the senior physicians had left the room, the younger medical students quietly removed the flowers and began using the instrument on one another. What they discovered left them awestruck.

Hodgkin's embrace of the stethoscope was emblematic of a larger shift in medicine—from intuition and external signs toward systematic, internal observation. He would later contribute to this transformation in his own right, identifying the lymphatic disease now known as *Hodgkin's lymphoma*.

The stethoscope marked the beginning of a seismic shift in medicine. No longer reliant solely on the visible and palpable, physicians could now explore the body's inner workings through sound.

Laennec's invention, and the science he built around it, culminated in his landmark treatise, *De l'Auscultation Médiate*. This comprehensive guide to chest examination through the stethoscope became required reading for a generation of European physicians.

Doctors travelled to Paris to witness the instrument in action. The stethoscope, once doubted, now hung from the necks of physicians across the continent.

René Laennec's domestic life took a brighter turn in December 1824 when he married Jacquette Argou. A widow and an old family friend, Jacquette had initially joined his household as a housekeeper, but over time, a deep affection had grown between them. Her quiet strength became an anchor for him.

Laennec, who had spent much of his life immersed in study and medicine, found in her a companion who understood him—his frailty, brilliance, and unwavering sense of purpose. Their days together brought him a joy he had long yearned for, a quiet contentment that softened the edges of his demanding routine.

Buoyed by this happiness, he published a new edition of his treatise and entered the competition for the Monthyon Prize in Physiology. His reputation, already formidable, continued to rise. Yet life's cruel balance soon reasserted itself.

Not long after their marriage, Jacquette suffered a miscarriage. The loss struck them both deeply. For Laennec—so intimately acquainted with illness, so often surrounded by death—this personal tragedy was a fresh wound. Here was a sorrow he could not diagnose, a life he could not save. The emotional weight bore down on him, and his health began to falter.

His chest symptoms returned: the familiar fever, the persistent coughing, the shortness of breath. By May 1826, it was clear he could no longer endure the punishing rhythm of Parisian life. The city, once a symbol of promise, had become too much. He returned to his childhood estate in Ploaré, Brittany, in search of rest and relief.

The countryside welcomed him with a gentleness that Paris could not offer. The soft winds carried the scent of the sea, the fields swayed with the rhythm of old familiarity, and the days passed more slowly. Jacquette remained at his side, her presence a balm. His nephew, Mériadec, visited often, eager to be near the man whose intellect he revered.

But Laennec knew his time was growing short. One day, weakened but lucid, he handed Mériadec the very instrument he had invented.

"Listen," he said softly, guiding his nephew's hand to his chest.

Mériadec hesitated. He understood all too well what he might hear. Still, he placed the stethoscope against his uncle's frail body. The sound echoed back unmistakably: the hollow resonance of cavitating tuberculosis.

His face fell. Laennec watched him with quiet intensity, reading the answer in his expression long before any words were spoken.

"So, it ends this way," Laennec murmured. There was no bitterness, only a calm acceptance. The irony was sharp but not cruel: the stethoscope, his greatest gift to medicine, had delivered the final diagnosis—of his own fate.

"I've been prepared for this," he said, his voice steady. "And I'm at peace."

The weeks that followed were slow and still. The man whose mind had once raced ahead of his time now moved with the rhythms of the land. He found solace in the cries of gulls overhead, in the soft rustle of wind through trees, in the distant crash of waves against the cliffs.

On 13th August 1826, René Laennec died at the age of 45. His final days were spent surrounded by those who loved him—Jacquette, ever faithful, by his side.

Though his name would later become associated with "*Laennec's cirrhosis*," it was the stethoscope that he regarded as *"the best part of my legacy."*

What is Tuberculosis?

Tuberculosis (TB) is an infectious disease caused by the bacterium *Mycobacterium tuberculosis*. It primarily affects the lungs but can also impact other body parts, such as the kidneys, spine, and brain.

In the 19th century, tuberculosis was a devastating and poorly understood illness, often referred to as "consumption" because of the way it wasted the body. It spread through crowded cities and poor living conditions, claiming millions of lives. At the time, it was mysterious, untreatable, and widely feared.

Today, we know that TB spreads through the air when a person with active tuberculosis of the lungs or throat coughs, speaks, or sneezes, releasing tiny droplets containing the bacteria. Symptoms of active TB include a persistent cough lasting more than three weeks, chest pain, coughing up blood, fatigue, weight loss, fever, and night sweats.

While TB can be serious and sometimes fatal, it is now treatable with a long course of antibiotics. Advances in medicine have transformed the fight against this ancient disease.

CHAPTER 8

A Stroke of Insight

THE SOUTH ATLANTIC stretched wide and tranquil under the morning sun, its gentle waves brushing the hull of the H.M.S. *Voltaire* as it cut through the clear blue waters. It was 1st April 1941; a date now etched in the midst of a war that had reshaped the world. France had fallen. The British had narrowly escaped Dunkirk. And now Hitler stood poised to crush England—his only remaining foe.

Against this backdrop of global turmoil, the *Voltaire*, a proud vessel of the British Royal Navy, steamed towards Freetown, Sierra Leone. There, it would lead a convoy of freighters through waters teeming with unseen threats. The whirring of the ship's engines blended with the sighs of the sea—a fragile harmony that masked the tension simmering just beneath the surface.

Below deck, Dr Charles Miller Fisher, the ship's lone Canadian doctor, lay back on his cot in the cramped quarters he called home. For the first time in days, he allowed himself a moment of stillness. His thoughts turned to his pregnant wife, Doris, far away in Canada. The memory of her light, lilting laugh surfaced in his mind, bringing a fleeting smile to his face. He closed his eyes, clutching the image tightly, willing himself to believe he would be there when their child took its first breath.

The blare of the alarm shattered the quiet. Fisher bolted upright, his chest tightening as the siren's shriek reverberated through the ship. This was no drill. Boots thundered above, orders were barked, and the familiar rush of adrenaline surged through his veins. He stumbled into his boots, heart pounding, and raced towards the medical bay.

A young ensign passed him in the corridor, face pale and lips trembling. "A German ship!" the boy gasped before vanishing into the chaos.

By the time Fisher reached his post, the crew had already sprung into action. The sick bay buzzed with frantic energy as assistants scrambled to prepare for the inevitable flood of casualties. He took stock of their limited supplies—splints, bandages, vials of morphine—wondering grimly whether it would be enough.

He didn't have to wonder for long.

The first barrage struck with bone-rattling force, shaking the ship to its core. A deafening explosion followed, and the sick bay erupted into chaos. A shell ripped through the metal walls, sending shrapnel flying and plunging the room into smoke and blood. Fisher's ears rang as he dragged a wounded sailor to a corner, blood slicking his hands. The smell of burning fuel and the copper tang of blood turned his stomach. His carefully laid plans for triage disintegrated, replaced by the raw calculus of survival. Tourniquets were tightened without ceremony; morphine was rationed with brutal precision.

Another explosion rocked the ship. A voice, barely audible over the din, cried out: "Abandon ship!"

Fisher hesitated only long enough to ensure the remaining patients were being evacuated before joining the stream of men scrambling to the deck. The *Voltaire* was listing badly now, its death knell groaning through the hull. As Fisher leapt into the sea, the cold embrace of the Atlantic stole his breath, driving Doris and the baby from his mind—if only for a moment.

Beneath the waves, the world was muffled and alien. But when Fisher surfaced, the cacophony of battle was gone. Around him, the ocean stretched indifferent, dotted with survivors clinging to debris.

For a moment, he floated in stunned silence, lungs heaving. A swell lifted him just high enough to glimpse the *Voltaire*, lying on her side like a wounded beast before she slipped beneath the water for good.

When a rope ladder from the sunken ship came within reach, he fastened one end to his waist and let the other dangle from his feet. The salt stung his eyes as he scanned the horizon, searching for anything—a ship, a lifeboat, any sign of hope. Around him, the sea heaved and sighed, uncaring.

Time blurred. The sun rose higher, its harsh glare scorching his exposed skin. Doris came to him again, her image sharper now. He imagined her cradling their child, her face lit with the soft glow of maternal love. He clung to that vision as tightly as to the rope beneath him.

Hours passed—or were they minutes? The rhythmic rise and fall of the waves lulled him into a strange calm. The notion of swimming—either back to South America or forward to Africa—crossed his mind, though the distance in either direction was equally absurd. So, he waited, conserving strength.

Then, on the crest of a swell, a shape appeared on the horizon. His heart leapt. He tried to wave an arm, but exhaustion pinned him to the water. The ship drew closer, its outline sharpening. Relief coursed through him—until the crack of machine gun fire shattered the air.

His rescuers were not allies.

The German vessel loomed, a predator drawing its prey from the sea. Rough hands hauled him aboard. He collapsed onto the deck, gasping, catching fragments of German: sharp words, barked commands. The grim truth settled in—he and the other survivors were to be taken back to Germany.

Though his body ached with the relief of escape from the ocean, a new dread settled over him. Captivity awaited. As he was dragged towards the ship's hold, he clenched his jaw and summoned Doris and the baby to the forefront of his mind.

They would keep him alive.

They *had* to.

The German vessel cut through the cold waters, its steel hull rising and falling with the waves. Fisher sat on the sun-scorched deck, his wrists bound and his back pressed against the unyielding metal rail. The heat radiated off the ship, the air shimmering with it. Around him, other survivors sat slumped in silence, their faces etched with exhaustion and dread. The sharp tang of salt mixed with the faint stench of oil—a bitter reminder of the battle that had ended their freedom.

Fisher stared out at the endless horizon, his mind retreating somewhere safer—far from the oppressive heat and the watchful eyes of their captors.

He thought of Doris. Her face rose unbidden, vivid and warm in his memory. He could almost hear her laugh—light, irrepressible, capable of lifting a room. He closed his eyes and held the image close. But as the ship rocked beneath him, the harshness of the present returned.

His thoughts drifted backwards.

Born in 1913 in the quiet town of Waterloo, Ontario, Fisher's early years were shaped by a close-knit community. His childhood was unremarkable in the best sense—cold winters, warm summers, family dinners, and small adventures along the Grand River. By the age of ten, his path seemed preordained. His parents often spoke with quiet pride of "our Charles, the future doctor," and Fisher never seriously questioned the role he was expected to grow into.

His school years passed in a blur of textbooks and sports fields. By the time he enrolled at *Victoria College, University of Toronto*, he was already known for his sharp mind and steady drive. The demanding seven-year programme, combining a Bachelor of Science with medical training, tested him—but Fisher thrived. Amid the grind of lectures and clinical rotations, he still found time for swimming and water polo, the water offering him a rare sense of freedom.

The memory of his graduation brought the faintest of smiles to his face. It had been a day of promise, of pride. His internship at *Henry Ford Hospital* in Detroit felt like the first true step towards a life of purpose. But it was at Montreal's *Royal Victoria Hospital* that his world truly changed.

There, in the busy wards and hurried corridors, he met Doris. Her wit was sharp, her kindness disarming. They were drawn together in quiet, unassuming moments—shared glances, late-night walks, conversations that stretched into dawn. Their courtship was brief, but emotionally rich. As war loomed, they exchanged vows in a quiet ceremony, promising each other a future neither could guarantee.

For six precious months, they lived that promise. Shared breakfasts, gentle arguments, a rhythm of ordinary days that felt extraordinary in retrospect.

But war waited for no one. When Canada entered the conflict in 1939, Fisher volunteered for the Canadian reserve militia. He felt it not just as duty—but necessity. When the Allies called for naval medical officers, he transferred to the British Royal Navy.

He left Doris with whispered assurances and the unborn life growing within her. His service aboard the *Voltaire* had been gruelling but meaningful, filled with camaraderie, purpose, and the ever-present urgency of medicine at sea. Until the *Thor*—the German ship that changed everything—rose on the horizon.

A sharp shout jolted him back to the present. The *Thor's* deck was no place for drifting thoughts. Fisher straightened, back aching from hours of sun and silence.

Ten days blurred past aboard the raider—ten days of hunger, heat, and humiliation. The guards barked orders with mechanical indifference, and the prisoners endured. When they were finally transferred to an old Dutch prison tanker, Fisher thought things couldn't possibly worsen.

He was wrong.

The tanker's after-hold was a pit. The air was thick with the reek of sweat, rot, and despair. Space was so tight that men slept sitting up, backs pressed against one another. By day, they were allowed brief stints on deck, the sea air like a forgotten blessing. At night, they were herded back below, the hatch slammed shut.

Fisher often lay awake in that suffocating dark, his thoughts swinging between the memory of Doris and the cold inevitability of what lay ahead.

When the ship reached La Rochelle, France, the prisoners were marched through the port beneath the watchful eyes of armed guards. The city wore its wounds openly—shattered windows, scorched walls, and an unnatural stillness. Their destination was a prison camp in St. Médard, where the conditions made the tanker seem almost merciful. The weak flour soup barely sustained them, and as more men arrived, overcrowding became oppressive.

Fisher's medical knowledge afforded him some utility. He treated what he could, offering more comfort than cure. Supplies were scant. But even a kind word, a gesture of care, reminded the other prisoners they hadn't been entirely forsaken.

When the camp reached its limit—400 men—they were packed into cattle cars for a journey Fisher would never forget.

The train jolted and groaned through a war-ravaged Europe. The air inside the cars grew foul within hours, nausea rising with the endless rocking. Stops were rare; but when they came, the heavy silence of the war-torn towns outside only deepened their despair.

Fisher clung to a single image: Doris, holding their child. Had the baby been born? Was she still waiting for word that he was alive?

When the train finally stopped at *Stalag XB* in Sandbostel, Germany, Fisher's legs nearly buckled beneath him. He stumbled from the car onto frozen earth. Around him, other men emerged, their faces pale, eyes hollowed.

He took a long breath and braced himself.

Whatever lay ahead, he would endure.

He *had* to.

The harsh expanse of *Stalag XB* stretched lifelessly across the sand dunes of northern Germany, a barren prison carved out of desolation. Not a blade of grass broke the monotony; no tree offered shade or respite from the wind that swept mercilessly over the camp. The compound, hemmed in by double barbed-wire fences, spanned a grim 200 by 400 yards. Guard towers loomed at precise intervals, their mounted machine guns and sweeping spotlights ensuring there was no escape from their gaze. Guards patrolled the perimeter with enormous dogs at their sides, and the crunch of boots on gravel was a constant reminder of captivity.

For Fisher, the bleak landscape mirrored the isolation within him. Once a doctor, a husband, a man with purpose—he was now a number among thousands, a prisoner in a sea of hollow faces.

Each day in *Stalag XB* blurred into the next, marked only by the rituals of survival. He thought often of Doris, and of the child he had never met. Somewhere in Canada, life continued—he imagined her walking through snow-covered streets, hand resting on her belly, eyes scanning the horizon for news. He didn't know that their daughter, Elizabeth, had entered the world on 2nd April 1941, just a day after the *Voltaire* was sunk.

Doris had learned of the ship's fate through a newspaper article read aloud by her father. The initial reports plunged her into anguish, and the naval telegram confirming Fisher as '*missing in action*' had only deepened her despair. It wasn't until late June, when word of

Fisher's captivity reached her family, that a fragile thread of hope began to emerge.

In the camp, Fisher began to scrape together a semblance of his former self. With ingenuity and persistence, he established a rudimentary medical setup using whatever he could barter or scavenge. The so-called "infirmary" was a cruel parody of a clinic: 12 narrow beds, paper bandages that tore with the slightest pull, liquorice tablets for gargling, and a single corroded scalpel. Sulpha drugs—the most effective antibiotics of the time—were virtually non-existent.

Still, Fisher worked tirelessly, treating injuries, managing fevers, offering not just medical aid, but something rarer: the sense that someone still cared. Some days, the placebo effect was all they had. He learned to disguise his helplessness with a calm voice and a steady gaze. "We'll get you patched up," he'd say, even when his hands held nothing but water and cloth.

Stalag XB became a strange kind of classroom. Fisher studied German, learning from overheard conversations, smuggled pages of books, scraps of paper passed like currency. He read whatever he could get his hands on—history, literature, navigation—turning the monotony of captivity into a defiant act of learning.

But the reality of camp life could not be romanticised. The prisoners' frustration with their confinement grew as the months dragged on. The desire for freedom burned hotter with every day spent behind the barbed wire. Then came the whispers: an escape plan.

A young naval officer had proposed digging a tunnel beneath the perimeter. At first, it sounded impossible. But desperation lent conviction. The plan spread quietly. Men volunteered in shifts, working at night, hiding their efforts beneath the barracks. They shored up the passage with wooden slats stolen from their own bunks, passing soil back in their trousers or stuffed into hollowed-out books. It was back-breaking, dangerous work, but it gave the prisoners something they hadn't felt in a long time: purpose.

Fisher knew he could not actively participate. As a medical officer, the *Geneva Convention* forbade him from engaging in escape attempts. Still, he watched. He offered quiet encouragement. He stitched up hands torn by wood splinters and kept his silence when guards lingered too long nearby.

Over four months, the tunnel took shape—a cramped artery of hope beneath the prison. It ran 100 yards, dug five to nine feet below the surface, just wide enough to crawl through. Its exit emerged in the surrounding peat bogs. Civilian clothes were sewn from scavenged cloth. Counterfeit papers were drawn by hand. Maps were memorised. German currency was gathered.

And somehow, the guards never found it.

The night of the escape crackled with tension. The prisoners staged a raucous variety show to distract the guards, their voices rising in song and laughter that felt almost real. While the guards' attention drifted, fourteen officers slipped into the tunnel one by one, vanishing beneath the earth.

Fisher watched from the shadows, his heart hammering in his chest. For the first time in months, he dared to believe something good might happen.

But the dream did not last.

By morning, grim news spread through the barracks: every escapee had been captured. One had made it as far as 125 miles south before being caught. The others had been intercepted within hours.

Fisher sat on his bunk, his head in his hands, as the weight of their failure settled over the camp. Rumours followed—that the men had been sent to high-security punishment camps, that executions had taken place. Nothing could be confirmed, but the mood shifted. The elation that had once hummed beneath the surface was gone.

And yet, Fisher could not bring himself to see the attempt as a failure. In his notebook, he later wrote:

"Is there a lesson in all of this? I see in it evidence that in every man a nobleness of spirit resides; can we but summon it forth. Silently, a common bond united us. We were fighting for right against a ruthless subhuman enemy plain to see. Churchill and the British people set the example."

There, among the bleakest days, that bond had meant everything.

The summer of 1942 brought a new chapter in Fisher's captivity. He was transferred to *Marlag* (Milag Nord) in Westertimke, a naval prisoner-of-war camp deeper in the German countryside. But any hope that conditions might improve was quickly extinguished.

The desolation of *Marlag* mirrored that of *Stalag XB*. The barren land stretched endlessly under an oppressive sky, devoid of life. Not a single tree or blade of grass softened the harshness of the landscape. Encased in barbed wire and watched by guards armed with rifles and rigid suspicion, the camp felt like a physical manifestation of despair.

Life in *Marlag* revolved around survival. The daily diet was meagre, monotonous, and cruelly insufficient. Black bread, turnips, sauerkraut, potatoes, and watery fish soup made up the staples, with occasional scraps of sugar, dried peas, or tinned fish appearing like rare luxuries. The lack of calories, protein, and essential nutrients slowly drained the men. They aged visibly—faces hollowed, ribs protruding, skin stretched tight over sharp bones.

Even minor injuries refused to heal. A small cut could become infected. A fever might turn fatal. Fisher watched helplessly as hunger thinned the men around him, and strength gave way to inertia. He worked tirelessly, but each day the scale of suffering grew.

Then came disease.

Shortly after his arrival, a diphtheria epidemic swept through the camp. The bacterial infection attacked the throat and lungs, leaving its victims gasping for air. The German authorities, indifferent to their suffering, provided only minimal doses of antitoxin. The outcome was devastating. Many died. Others survived but were left paralysed.

Fisher faced each day in the infirmary with dread. The medical supplies were few, and his hands often trembled as he measured out the precious antitoxin—droplet by droplet—knowing it might not be enough to save the man lying before him. He slept little. He endured much.

When the diphtheria finally ebbed, it was replaced by bloody dysentery (a severe intestinal infection causing diarrhoea with blood). The infection spread rapidly through the weakened camp population, leaving men doubled over in pain, dangerously dehydrated.

Fisher's frustration boiled over during a particularly desperate exchange with the German camp doctor—a man whose disdain for sulphonamides, the life-saving medication, bordered on contempt.

"I need more sulpha," Fisher insisted, his voice tight.

The doctor sneered. "From which university did you graduate?"

"Toronto," Fisher replied, keeping his voice even.

The doctor gave a dismissive snort. "That is in America. Pah! You are no doctor. When students in Europe are too 'dumb' to pass their examinations, they go to America and get their degree."

Fisher said nothing. He had learned that survival sometimes required silence. But inside, he burned with anger. Not for the insult—he had endured worse—but for the men who would die because of another man's arrogance.

Then came winter, ushering in suffering of another kind.

Hundreds of starving Russian prisoners were moved into a neighbouring compound. Their skeletal forms and hollow eyes made even the emaciated British prisoners seem well-fed by comparison. With them came another threat—typhus.

The disease swept through the Russian camp like wildfire. Bodies were carted away daily. Groans and cries echoed into the night.

Fisher and his fellow prisoners acted swiftly. They isolated their section of the camp, established inspections, and enforced hygiene protocols with near-military discipline. They spread fearful rumours—tales of horrific punishments for non-compliance—not out of cruelty, but as a last-ditch effort to prevent the disease's spread.

The strategy worked. The British prisoners were spared. But at night, Fisher lay awake in his bunk, listening to the muffled suffering just yards away, helpless to intervene.

And yet, even amid all this, he found ways to endure.

His growing fluency in German granted him access to medical texts, and he devoured everything he could find. He scoured smuggled scraps, pored over outdated monographs, anything that might further his knowledge.

By September 1944, after more than three years in captivity, a flicker of hope finally pierced the fog. Fisher was among the few prisoners selected for repatriation—part of a rare exchange of wounded and medically vulnerable prisoners of war.

The announcement barely registered at first. It felt surreal. After everything—after hunger, disease, diphtheria, dysentery, typhus, and the relentless erosion of body and spirit—he was being sent home.

Even as the arrangements unfolded, Fisher struggled to believe it. He had learned not to trust fortune, not to lean too heavily on hope. But as the days passed, the plan solidified. He would soon leave the camp. He would see Canada again.

And Doris. And the child.

Was it a boy or a girl? Did they think of him, as he had thought of them every day since he had been taken? The anticipation of seeing his wife again, of holding his child for the first time, became a beacon that pulled him through those final weeks. He had survived, and soon, he would be free.

The reunion was everything he had imagined—and nothing he could have prepared for.

When he stepped off the transport, the crowd of families waiting at the station blurred before his eyes. Then he saw her. Doris. Standing still, apart from the bustle, her hands clasped around a small one. Her eyes locked on his.

She hadn't changed—not really. Her frame was slimmer, the shadows beneath her eyes a little deeper, but she was still his Doris. And she was here.

"Charles!" Her voice rose above the noise, breaking like sunlight through overcast skies.

He dropped his bag.

In an instant, she was in his arms. He held her as tightly as he dared, afraid she might vanish, afraid he might wake up. Tears streamed freely. The crowd disappeared. Time dissolved.

He buried his face in her hair, breathing in the scent that had lived only in memory for so long.

When at last she pulled back, her hands gripped his arms, still disbelieving. His gaze shifted to the child at her side.

Elizabeth.

She wasn't the baby he'd imagined during those long years, but a toddler now—wide-eyed, uncertain, her soft curls framing a face both unfamiliar and heartbreakingly familiar.

He dropped to one knee. His voice trembled. "Hello, Elizabeth."

She stared at him, clutching a small toy. Her brow furrowed, curious and cautious.

Then, slowly, she stepped forward.

Her tiny hand reached out, brushed the stubble on his cheek. Fisher held his breath.

Then—she smiled. A wide, fearless grin. And without hesitation, she flung her arms around his neck.

His breath caught. He held her tightly, tears falling in silence.

Doris knelt beside them, wrapping her arms around them both.

And for the first time in years, Fisher felt whole.

The days that followed were tender, surreal, and quietly miraculous.

The smallest things—Elizabeth's laughter, Doris's touch, the smell of home-cooked food—felt like treasures. Fisher revelled in the simple rituals of home life. Breakfasts, bedtime stories, soft footsteps across the kitchen floor. These were not luxuries—they were redemptions.

But the stillness couldn't last forever.

Even as he cherished every moment, a quiet restlessness stirred within him. The war had taken years he couldn't get back. Years he should have spent honing his craft. He had survived—but now he needed to rebuild.

The Canadian government, recognising that his captivity had denied him specialised training, arranged for Fisher to resume his medical career at the *Royal Victoria Hospital.*

He returned to Montreal with a sense of mission. He threw himself into his studies, first exploring endocrinology and diabetes. These fields held promise, and he felt drawn to their complexity. But it was a two-month rotation in neurology at the *Montreal Neurological Institute* (MNI) that changed everything.

The brain. It captured his imagination in a way nothing else had. Its mysterious workings, the delicate architecture of thought, sensation, identity—it was a puzzle he felt born to solve.

Then came the case that sealed it.

A United States Army general arrived at the MNI with a peculiar complaint: just before losing consciousness, he would hear the rhythmic beat of tom-toms.

The symptom struck Fisher as oddly specific. Intrigued, he delved into the literature, cross-referenced cases, chased down obscure reports. The pattern that emerged pointed to a lesion in *Heschl's gyrus*—the region of the brain responsible for processing sound.

His diagnosis proved correct.

Dr Wilder Penfield, the eminent neurosurgeon and director of the MNI, took notice. After the diagnosis was confirmed, he approached Fisher.

"Have you ever considered neurology?"

Fisher hesitated. "I haven't. Not seriously."

Penfield smiled, just faintly. "You should. We could use someone like you."

And just like that, a new path opened.

Penfield offered him the position of acting registrar. Fisher accepted. And from that moment, he never looked back.

After two formative years at the *Montreal Neurological Institute*, Fisher's tireless dedication once again caught Penfield's attention. Recognising his aptitude—and the depth of his curiosity—Penfield encouraged him to pursue advanced training in a field still barely defined: cerebrovascular disease.

It was an area with more questions than answers. But that suited Fisher. He wasn't looking for easy.

With Penfield's recommendation, Fisher secured a place under Dr Raymond D. Adams at *Boston City Hospital*, a leader in the study of stroke and brain injury. In January 1949, he moved to Boston to begin a year-long fellowship.

It was here that everything began to coalesce.

One ordinary afternoon in the pathology lab—formaldehyde

thick in the air—Fisher was dissecting the brains of ten recently deceased patients. Each case demanded his full attention. Each brain told its own story.

But as he examined them, something began to nag at him. There was a pattern.

Several of the brains showed small patchy areas of haemorrhage (bleeding) within regions of damaged brain tissue caused by a lack of blood flow - but without any clots or obstructions in the major arteries that might have explained them. Initially, the findings seemed inconclusive. Curious, but not extraordinary.

But Fisher couldn't let it go.

He returned to the patients' medical records and discovered a link: three of them had atrial fibrillation.

Atrial fibrillation wasn't unfamiliar—it was known to cause an erratic heartbeat, disrupting the smooth flow of blood through the heart. But Fisher began to suspect something deeper. He wondered: could these irregular rhythms be generating clots that travelled to the brain?

And more than that—could these clots, once lodged in small cerebral vessels, *first* cause a blockage, and *then*, when they broke down, trigger the surrounding tissue to haemorrhage?

It was a theory that challenged the prevailing view. At the time, most brain haemorrhages were attributed to longstanding high blood pressure, which was thought to weaken small penetrating arteries, leading them to rupture. But Fisher's observations pointed in another direction: *embolism.* These clots, originating in the heart due to atrial fibrillation, could travel through the bloodstream and lodge in smaller blood vessels in the brain, causing blockages. Over time, the ischemic tissue—the area deprived of oxygen—became fragile. When the clots broke apart, and blood flow was restored, the weakened capillaries and small blood vessels were prone to rupture, resulting in a phenomenon known as *haemorrhagic transformation.*

That phrase—*haemorrhagic transformation*—would later become central to stroke research. But in 1949, it was just a hunch, based on careful dissection, cross-referenced data, and the stubborn instinct that something important had been overlooked.

Fisher didn't wait to be invited into the conversation. He submitted his findings to leading pathology journals.

He was rejected.

The rejections didn't surprise him. The medical establishment was slow to let go of familiar narratives. Challenging them—especially without a senior academic name attached—was unwelcome.

But Fisher didn't give up.

He kept digging—sometimes literally. He began spending long hours with old German medical literature, tracing ideas back through time, seeking not just answers, but *gaps*—places where curiosity had once sparked, then gone quiet.

Rejection wasn't failure. It was fuel.

In 1950, Fisher returned to *Montreal General Hospital*, eager to continue his work. It wasn't long before he encountered another case that would redefine his understanding of stroke. At the *Veterans Hospital*, a patient described symptoms that caught Fisher's attention.

"Doctor," the man said, his voice steady but troubled, "before I became paralysed, something strange kept happening. I would go blind in one eye, just for a few minutes. Then it would pass. But when the paralysis came, it hit the opposite side of my body."

Fisher's mind clicked into gear.

"The blindness," he asked, "was it always in the same eye?"

The patient nodded. "Always."

It was a crucial clue.

What the man described wasn't just coincidence—it was a warning. The transient blindness, or *amaurosis fugax*, suggested brief interruptions in blood flow to the eye. Fisher suspected a narrowing of the carotid artery (one of the major blood vessels in the neck that supplies blood to the brain) on the same side.

And when the stroke occurred, it affected the *opposite* side of the body—because the brain is wired so that one hemisphere controls the other.

This pattern—temporary, reversible symptoms that preceded a major stroke—had not yet been named.

Fisher gave it one: *transient ischaemic attacks*, or TIAs.

The case stayed with him. So did the unanswered questions.

When the patient later died of metastatic cancer, Fisher learned—too late—that no autopsy had been performed. But he couldn't let the trail go cold. He reached out to the family and obtained permission to conduct a post-mortem himself.

Late at night, in a quiet funeral home, Fisher worked under sterile light, the smell of embalming fluid thick around him. With calm, methodical hands, he examined the carotid arteries.

There it was. A blockage.

A narrowed segment of the artery, partially obstructed, precisely where he had expected it to be.

It was the evidence he needed. The transient blindness hadn't been a mystery symptom. It had been a *signal*. A chance to intervene before catastrophe.

Fisher began collecting more data. Over time, he examined 1,100 pairs of carotid arteries, building an irrefutable case. These blockages were not rare. They were a leading cause of stroke.

The implications were seismic. If doctors could identify and treat these blockages—especially in patients with TIAs—they could prevent full-blown strokes.

This research laid the groundwork for a surgical procedure that would become lifesaving: carotid endarterectomy (a procedure to remove plaque buildup from the carotid arteries).

In 1954, Dr Raymond D. Adams—now Chief of Neurology at *Massachusetts General Hospital*—invited Fisher to return to Boston to develop the hospital's first dedicated stroke service.

It was a defining moment—not just in Fisher's career, but in the field itself. Stroke care, long neglected and fragmented, was now becoming a specialised discipline.

Under Fisher's guidance, the stroke service grew into a world-leading programme. Trainees came from around the globe. Some would go on to build stroke centres of their own, spreading his influence far beyond Boston.

Fisher had helped elevate stroke from the margins of neurology into a central, urgent field of study.

Fisher's dedication to his work was evident in every aspect of his career. His rounds were thorough and exhaustive, aimed at extracting every possible detail from each patient. He believed that retrieving small but critically relevant details separated the expert from the novice.

This approach sometimes frustrated those looking for quick answers, but it earned him respect from those who understood the gravity of his work. To Fisher, medicine was about truly understanding his patients and giving them the dignity of his full attention.

Fisher had a unique habit of categorising unusual patient signs and symptoms. His notes were as thorough as his rounds, with carefully labelled folders for cases that intrigued him:
- "patients who wrote off the paper"
- "mumblers"
- "irascible patients"
- "topplers"
- "pure sensory stroke"

He saw patterns where others saw chaos; in those patterns, he found the keys to diagnosing and treating neurological diseases.

His professional life was complemented by the unwavering support of his wife, Doris. While Fisher often focused on his work, Doris created the stability he needed to excel. She fielded late-night phone calls, handled the logistics of his demanding schedule, and drove to the hospital to collect him after gruelling days.

Yet their relationship was far more than one of practicality. Fisher was not known for outward displays of emotion. He rarely spoke of the past. But in Doris's presence, something softened. His stress lifted, replaced by a quiet ease from knowing he had someone who understood him completely. Doris provided Fisher with more than just love; she gave him balance. Her grounding influence allowed him to fully immerse himself in his work, knowing he had a partner who would always support him.

Fisher's legacy would continue to expand in vocabulary as well as practice. He coined terms that today are commonplace in neurology:

- *Transient monocular blindness*
- *Subclavian steal*
- *Symptomatic normal pressure hydrocephalus*
- *The string sign*
- *Transient global amnesia*

Each term emerged not from theory, but from observation. From a lifetime of watching the body speak in fragments and finding the language to decode it.

One of the most notable conditions associated with his name is *Miller-Fisher syndrome*.

On his rounds, Fisher began noticing patients with unique symptoms. These patients shared three features:

- Difficulty coordinating their movements (ataxia)
- A loss of reflexes (areflexia)

- Paralysis of the eye muscles (ophthalmoplegia), often
  leading to double vision

They weren't weak in the usual sense. Their muscles worked—but the control, the communication, was off.

Fisher identified a consistent pattern that had previously gone unrecognised. His ability to connect these symptoms led to the recognition of *Miller-Fisher syndrome* as a separate condition, providing clarity and a new framework for diagnosing this new neurological disorder.

Fisher was not just a clinician; he was a scholar. He spent long hours in the Harvard library, reading English and German medical texts, sometimes from centuries past.

"We cannot afford," he said once, "to redo the history of neurology every 20 years."

Fisher didn't seek fame. But he welcomed discussion. His door was always open—to students, to colleagues, to anyone hungry for knowledge. He was a man who believed that medicine, at its best, was a conversation.

Even as he approached the age of 96, Fisher's passion for neurology remained undiminished. He continued publishing journal articles, his insatiable curiosity driving him to contribute to the field he loved until the end. Curiosity was not something he retired from.

In 2012, at the age of 98, Fisher passed away with Doris by his side. Her presence in those final moments brought him peace, just as it had during the most challenging periods of his life. She had been his source of hope as he floated in the cold Atlantic waters after the *Voltaire* was sunk, his anchor during the gruelling years in Nazi prisoner-of-war camps, and his partner through decades of discovery and dedication.

Her love was more than a source of strength. It helped him rise, again and again, into the man he became.

What is Miller-Fisher Syndrome?

Miller-Fisher Syndrome is a rare neurological condition that affects the body's peripheral nervous system. It is characterised by three key symptoms: loss of co-ordination, absence of reflexes, and paralysis of the eye muscles, which can lead to double vision and difficulty with balance and movement. The syndrome typically develops after a viral infection and is classified as an autoimmune disorder, where the body's immune system mistakenly attacks its own nerves. Diagnosis often involves clinical examination, alongside testing for specific antibodies associated with the condition. While most patients recover fully with treatment, such as immunotherapy, early diagnosis is critical to managing symptoms effectively.

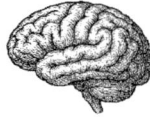

CHAPTER 9

Forgotten

IN THE HEART of Imperial Germany, a nation brimming with ambition, Auguste Deter was born in 1850, in the thriving city of Cassel. It was a land in flux, shedding its agrarian skin as the roar of machinery began to dominate towns and cities. With the unification of the German states under Prussian rule, the political map was redrawn—and so too the skyline, now threaded with railways and factory chimneys. Yet amid all this progress, tradition held fast. Women, especially, were still expected to find their purpose at home, their lives neatly framed by marriage, motherhood, and domestic duty.

Auguste was no stranger to hardship. By the time she was 14, childhood had been folded away like a worn apron. She found work as a seamstress's assistant, spending long hours hunched over fabric, stitching each piece with quiet precision. The work was exacting, but it suited her. Even then, she was diligent and unassuming, quietly proud of the skills she was learning. After nearly a decade of perfecting her craft, life turned a gentler corner. On a spring day in 1873, Auguste married Carl Deter—a kind, steady railway clerk with soft manners and steady eyes. Together, they left Cassel for Frankfurt, a city alive with promise. Carl's job held the hope of a modest but comfortable life, and they chased it together.

Their marriage settled into an easy rhythm of shared effort and quiet joy. Auguste devoted herself to their home, and what they built together was more than just four walls—it was a sanctuary. The scent of fresh-baked bread drifted through rooms filled with the sound of Carl's stories from work. When their daughter Thekla arrived two years later, Auguste's days took on new purpose. She doted on the little girl, teaching her the value of order and simplicity. Carl, ever dependable, adored his daughter and watched his wife with a quiet sort of awe.

Theirs was not a love of grand gestures, but of small, enduring acts—shirts folded just so, a favourite pudding waiting on the table after a long day. Ordinary, perhaps, but never empty. Their neighbours often commented on the peacefulness of the household, its warmth, its calm. It was a house that felt loved.

Then, slowly, the familiar began to slip.

In the spring of 1901, Carl noticed little things—barely anything at first. Auguste forgot a grocery item. Misplaced a recipe card. Nothing unusual. But the lapses kept happening, and soon, there was something else. One evening, as Carl walked through the front door, Auguste turned to him with eyes filled not with warmth but suspicion.

"Where were you?" she demanded. "I know you've been with that woman next door. Don't lie to me."

Carl froze. In all their years together, trust had never wavered. The words struck hard, not just for their accusation but because they came from her.

In the weeks that followed, her behaviour became increasingly strange. She tucked silverware into the linen closet, dropped keys into flowerpots, and scorched dishes she had once cooked flawlessly. One night, Carl found her drifting through the hallway in her nightgown, whispering to herself, the words a tangle of nonsense. The house no longer felt like their home. It was as if someone had taken their life and tilted it slightly, just enough to unsettle everything.

By autumn, the change in her was undeniable. She rang neighbours' doorbells at odd hours, slammed doors, snapped at shadows.

Carl, desperate to make sense of it, wrestled with a growing help-lessness. How could the woman he loved—the woman he had built his life with—be slipping away before his eyes?

By November 1901, the situation reached a breaking point. After months of sleepless nights and growing desperation, Carl turned to their family doctor, who listened intently as he described his wife's decline. The doctor's expression darkened. Recognising the severity of Auguste's condition, the doctor wrote a referral letter:

> *"Mrs Auguste D. . . has been suffering for a long time from weakening of memory, persecution mania, sleeplessness, restlessness. She is unable to perform any physical or mental work. Her condition needs treatment from the local mental institution."*

With this grim diagnosis, it became clear that Auguste needed more specialised care than Carl could provide at home. This wasn't the life he had promised her on their wedding day. But he knew he had no choice.

On a grey November morning, Auguste entered the *Municipal Asylum for the Insane and Epileptic* in Frankfurt am Main. She was trembling, smaller somehow than she had ever been. Once graceful, with eyes that had sparkled with wit, she now flinched at every hand laid gently on her arm. The staff tried to soothe her, but her fear twisted every kindness into something sinister. The stark white corridors of the asylum stood in sharp contrast to the cosy home she had once ruled with such quiet command.

Dr Nitsche, the junior physician handling her admission, was immediately struck by her case. Auguste was just 51—far too young to be showing such signs. Her symptoms didn't follow any script he recognised. It was more than memory loss. Her mind was unravel-ling, pulled into a knotted mess of confusion, suspicion, and dread.

He decided to bring the case to the attention of his senior colleague, Dr Alois Alzheimer.

In June 1864, in the quiet Bavarian town of Marktbreit, where the Main River flowed gently through winding cobbled streets, Alois Alzheimer was born to Anna Johanna Barbara Sabina and Eduard Román Alzheimer. His father worked in the office of a notary public, and his mother, devout and disciplined, instilled a deep Catholic faith in the family. Seeking better prospects, the Alzheimers moved to Aschaffenburg while Alois was still a boy. It was there, at the *Royal Humanistic Gymnasium*, that he first revealed the depth of his intellect. The school's demanding curriculum nurtured not only his aptitude for science but also his love of the humanities—seeds of curiosity that would take root and flourish in the years to come.

By the time he graduated in 1883, Germany was transforming at breathtaking speed. Industrialisation redrew the map of daily life. Railways stitched the countryside together, and scientific discovery seemed poised to explain everything. This spirit of progress was infectious. When Alzheimer enrolled in medical school, he did so with more than a desire to treat the body—he wanted to understand the enigmas of the human mind.

His studies took him across some of the most prestigious medical faculties of the day: Berlin, Tübingen, Würzburg. Each offered a new lens through which to view medicine. In Berlin, he attended the lectures of Carl Westphal, a pioneer in cerebral disorders and forensic psychiatry. Westphal's ideas—that mental illness had roots in the brain's physiology, and that patients deserved humane treatment—greatly shaped Alzheimer's thinking. They challenged the old assumptions and lit a quiet fire in the young doctor.

Though earnest in his studies, Alzheimer was not without his youthful indulgences. In his final year at university, he joined a fencing fraternity, throwing himself into its rituals with vigour. The social life was lively, sometimes too much so. On one memorable occasion, he paid a fine after he and his teammates caused a late-night commotion outside a police station. A fencing duel in the summer of 1884 left him with a disfiguring scar along the left side of his face—a wound he carried for life, and one that made him favour being photographed from his right.

He completed his medical training in 1887, earning his diploma from the *University of Würzburg*. That December, he began his first post as an assistant house officer at the *Municipal Asylum for the Mentally Ill and Epileptics* in Frankfurt am Main. The asylum, under the directorship of Emil Sioli, stood out for its progressive stance: a rare place where research and patient dignity coexisted. Here, Alzheimer's compassion found structure, and his clinical curiosity found a home.

One of the defining moments in these early years came in 1889, when he met Franz Nissl—a brilliant young neuropathologist whose work was quietly revolutionising the field. Nissl had developed a novel staining method that made the microscopic structure of brain cells visible in a way no one had seen before. His *"Nissl stains"* brought the nervous system into sharp, almost intimate focus.

What began as a professional encounter quickly evolved into a friendship of rare depth. The two men spent countless hours together—at work, in the lab, over meals—driven by the same insatiable questions. Nissl's influence ran deep. He challenged Alzheimer's assumptions, introduced him to new methods, and pushed him to think not just clinically but microscopically.

Their collaboration began to bear fruit. They investigated everything from general paresis (a neuropsychiatric complication of late-stage syphilis) to cortical atrophy (the wasting away of brain tissue), mapping out how structural changes in the brain could manifest as psychiatric illness. It was groundbreaking work. The cerebral cortex, once a mysterious frontier, was starting to reveal its secrets—and Alzheimer was there, peering through the lens.

In 1895, his personal life took a brighter turn. He married Cecilia Geisenheimer, a widow from a wealthy diamond-dealing family. Their marriage was warm and affectionate, a true partnership of mind and heart. Nissl stood proudly beside him as a witness at the ceremony. Over the next five years, Alzheimer and Cecilia had three children. Their home, filled with music and laughter, offered a sense of grounding he hadn't known before. Thanks to Cecilia's financial security, Alzheimer could dedicate himself fully to his work, free from the worry of income. With a growing family and a flourishing career, he began to dream of leading his own psychiatric hospital— where he could unite clinical care with the kind of research he believed could change lives.

But change, as ever, came unbidden.

In 1896, Nissl accepted a prestigious post at the *University Hospital of Heidelberg*, joining forces with Emil Kraepelin—one of the towering figures of psychiatry. It was a proud moment, but a bittersweet one. The close-knit rhythm Alzheimer had shared with his colleague was gone. Though they stayed in touch, exchanged letters and visits, the daily exchange of ideas had ended. For Alzheimer, it marked the close of a chapter.

Then, in 1901, tragedy struck. Cecilia died suddenly, only months after their third child was born. She was just 41.

Alzheimer was devastated. The home that had once been a place of joy now echoed with absence. He was a widower with three young children, hollowed out by grief. His sister moved to Frankfurt to help manage the household, but the wound was deep. Bereft, Alzheimer did what many grieving people do—he turned to work.

Not to escape, but to find something solid amid the ache. He poured himself into his clinical practice and his research with even greater intensity, searching for meaning in the only place he knew he might find it.

Despite the sorrow, his professional life flourished. By the end of 1901, he had become one of the asylum's most trusted physicians— careful, empathetic, and known for his keen eye in both the clinic and the lab.

On the morning of 26th November 1901, his junior colleague Dr Nitsche approached him with news of a new patient. A woman had been admitted the previous week—confused, frightened, her memory disintegrating by the day.

Her name was Auguste Deter.

That morning, Dr Alois Alzheimer walked the dim corridors of the asylum, his footsteps echoing faintly off stone floors and high, whitewashed ceilings The air was heavy with the muffled sounds of distant footsteps, quiet murmurs, and the occasional sharp cry from a patient. His mind, as usual, was preoccupied with the endless stream of cases that came through the institution. But this patient— this Auguste Deter—nagged at the edge of his curiosity.

There had been something in Nitsche's account, something that didn't fit.

He paused outside her room, hand on the doorknob, and drew a slow breath. Then he stepped inside.

The space was stark—just a bed, a small table, a single wooden chair. Auguste sat perched on the edge of the bed, twisting the hem of her skirt between restless fingers. A shaft of morning light cut across her face, revealing wide, fearful eyes. She looked at him, but there was no flicker of recognition. Just wariness. Suspicion.

"Good morning, Mrs Deter," he said, lowering himself into the chair beside her. His tone was measured, professional, but softened by warmth. "I'm Dr Alzheimer. I'd like to ask you a few questions."

She nodded slightly. Her brow creased. A silent unease passed between them.

"What is your name?" he asked.

"Auguste," she replied, her voice barely audible. It trembled on the edge of certainty.

"And your last name?"

"Auguste," she repeated, more uncertain this time. Her eyes searched his, as if for reassurance.

He paused. "What is your husband's name?"

"Auguste, I think," she murmured.

"Your husband?" he prompted gently.

"Ah... my husband," she echoed, the word slipping out as though it were unfamiliar.

The disorientation hung in the room like fog. Alzheimer leaned back, watching her with quiet intensity. Her mind seemed to reach and retreat in equal measure—grasping for answers, then slipping away.

He continued. Producing a pencil from his pocket, he held it up. "What is this?"

She hesitated. "A pen."

Other objects she named correctly—a purse, a key, a diary—but even so, her grasp on reality seemed tenuous, fragile.

Later, at lunch, cauliflower and pork were served. Auguste insisted it was spinach. As she chewed the meat, she claimed she was eating potatoes and horseradish. Alzheimer noted it all.

By evening, her speech was fragmenting. She stumbled through sentences, repeated phrases, substituted nonsense for meaning. When asked to write her name, she managed just a shaky "Mrs" before stopping. Her hand trembled over the page, paralysed by a mind that no longer knew where to go.

Alzheimer observed quietly, disturbed by the precision of her deterioration. What could cause such a profound decline in someone so young? Her cognition was unravelling with terrifying specificity.

The days that followed brought more questions than answers. Each morning, Alzheimer resumed his gentle interrogation.

"What year is it?"

"1800," she said, without hesitation.

"Are you ill?"

"Second month," she answered, the words unconnected to the question.

"What colour is snow?"

"White."

"Soot?"

"Black."

"The sky?"

"Blue."

"Meadows?"

"Green."

He watched the flickers of clarity dance behind her eyes, only to vanish moments later. Simple questions sometimes grounded her— but anything more complex sent her spiralling.

"If you buy six eggs at seven dimes each, how much is it?" he asked one morning.

"Differently," she said. Her face clouded with confusion.

"What did I just ask you?"

She looked up, almost plaintively. "Well, this is Frankfurt am Main."

The leap in logic startled him. She trembled. Her voice thinned.

"So anxious," she whispered, clutching her arms. "So anxious."

During a physical examination, she initially complied. Then panic flared. She recoiled.

"I will not be cut!" she cried, her voice rising. "I do not cut myself!"

Her fear was sharp and raw. Alzheimer stopped the exam, shaken by her distress. Whatever this was, it wasn't just memory loss. It was a soul slowly retreating from the world.

As the weeks passed, the depth of her decline became achingly clear. She wandered the ward, touching the faces of other patients, searching for something—someone—familiar. Often, those gestures were met with aggression. Her agitation would rise, and the staff would struggle to calm her.

She accused them of cruelty. Of wanting to harm her. Her paranoia growing more pronounced.

And yet, in rare moments—brief and almost ghostly—she softened. She would speak politely to staff, offering them imaginary tea, welcoming them as though they were guests in her parlour.

It was, for a heartbeat, a glimpse of the woman she had once been.

For most of her days, she found solace only in the bath. The warm water seemed to soften her restlessness, to wrap her in something familiar when everything else had grown strange. The staff noticed it—the way her breathing slowed, her body uncoiled, her expression momentarily calm. But the nights were another matter entirely.

After sunset, her fear swelled. Shadows became threats. Sounds twisted into imagined dangers. Often, she had to be confined to an isolation room—not as punishment, but to protect her from herself and from others. She would cry out in the dark, caught between memories that no longer belonged to her and fears she couldn't name.

And still, Carl Deter came.

Despite the toll, despite the rising costs, he visited as often as he could. The asylum fees weighed heavily, but he never faltered in his loyalty. He could have left. Many husbands did—abandoning wives institutionalised by illness, letting go of vows no longer convenient. But not Carl. He searched tirelessly for alternative care, for somewhere less expensive. Even then, he refused to consider divorce.

Alzheimer noticed. A widower himself, he understood the quiet heroism of love that stays when the easy exits have vanished. Carl's presence moved him.

When Carl confided in him—his worries, the pressure, the ache of watching Auguste fade—Alzheimer made a generous offer. He believed her condition held significance far beyond the personal. In exchange for Carl's permission to use Auguste's medical records, and to study her brain after her death, Alzheimer promised she could remain under his care, free of charge.

It was a heavy decision. A painful one. But Carl understood what was at stake. And with a heart still bound to the woman she had once been, he agreed.

Auguste Deter, 1902: Captured in the asylum, Auguste appears worn and dishevelled, her features swarthy and sorrowful. Dressed in the institution's nightshirt, she is deprived of the modest attire befitting a lady of her era, symbolising the significant decline and suffering she endured.

Through that winter and into the following year, Alzheimer's own career was shifting. His meticulous observations and clinical insight had earned growing respect. In the summer of 1902, Emil Kraepelin—the leading figure in German psychiatry—invited him to join a research institute in Heidelberg. It was the kind of offer that could reshape a career. But Alzheimer hesitated. He had long dreamed of becoming the director of a *Hessian State Hospital*—independent, in charge, able to unite research and care under one roof.

So he waited. Applied. Hoped.

When word came that his application had been rejected, he felt the sting of a door closing. It was hard not to wonder if he had let something slip through his fingers.

But fortune was not quite finished with him.

Franz Nissl, now working with Kraepelin in Heidelberg, stepped in once more. He urged Kraepelin to renew the invitation—insisting that Alzheimer's brilliance was too valuable to be left behind. The offer came again, and this time, Alzheimer accepted. With gratitude, and no small measure of relief, he prepared to move.

By the end of 1902, he had left Frankfurt for Heidelberg. A year later, he relocated again, this time following Kraepelin to Munich. There, at the heart of a pioneering research environment, Alzheimer found the tools and freedom he had long craved. But leaving wasn't easy. He had spent years in Frankfurt. The patients he had known—their stories, their suffering—remained etched in his memory.

Before his departure, Emil Sioli, director of the Frankfurt asylum, made a promise: he would keep Alzheimer informed about the patients whose cases had been of special interest.

Chief among them was Auguste Deter.

The woman who had walked into his life confused and afraid now haunted his thoughts—her words, her gaze, the slow unravelling of her mind.

Over the years, as Alzheimer advanced in his career, Auguste's condition deteriorated rapidly. The confused woman he had once spoken to, coaxing fragmented answers in the soft light of a ward room, was now unrecognisable. Her speech, already broken, dissolved into unintelligible murmurs. Her once expressive face had gone blank, emptied of personality, of light, of the woman she had been.

In her final year, she withdrew entirely. Curled into herself in bed, legs drawn up tightly against her chest, she barely moved. The moments of paranoia and agitation that had once defined her days gave way to an eerie stillness. It was not peace, but something bleaker. A kind of quiet erasure.

The clinical notes from this period captured the stark reality of her decline:

> *"Completely stupefied, always lying in bed with legs drawn up. Regularly soiled with urine and faeces: never says anything, mutters to herself, has to be fed."*

They captured the devastation of her decline. But even these grim observations could not fully contain the sorrow of it—the loss of dignity, of memory, of identity itself.

By the spring of 1906, her body was as frail as her mind. Pneumonia set in, and the bedsores that had long plagued her became infected. Septicaemia followed. Her weakened system had nothing left to fight with.

On 8th April 1906, at quarter to six in the morning, Auguste Deter died. She was just weeks shy of her 56th birthday.

Her death passed quietly. One more fading life in an institution full of them. There were no announcements, no public mourning. Just the final entry in a file.

But news of her passing reached Munich.

When Alzheimer read the report, a heaviness settled in his chest. He thought of her as she had once been—the fear in her eyes, the questions she could not answer, the rare flickers of humour or grace. Her suffering had left a mark on him. He could not let it end with a file closed and a name forgotten.

There was something in her case—something new. Something that did not fit the usual moulds of senility or psychosis. He needed to understand it.

He arranged for her medical records and her body to be sent to him in Munich.

When it arrived, Alzheimer approached the task not as a detached pathologist, but as a man carrying a solemn promise. Alongside two Italian physicians, he began the post-mortem examination. He worked carefully, methodically—his hands sure, his focus unwavering. He had performed countless autopsies. But this one felt different.

He prepared Auguste's brain for microscopic analysis, using the techniques he had refined during years of collaboration with Franz Nissl. Slice by slice, stain by stain, he peered into the tissue, seeking clues—something to explain what had happened to her.

And then, he saw it.

Under the microscope, strange clusters appeared—dense clumps of protein fragments wedged between nerve cells. Amyloid plaques. They choked the fragile architecture of the brain, disrupting the flow of information between neurons.

Deeper still, another anomaly emerged. Inside the nerve cells, twisted strands of protein—neurofibrillary tangles—like knotted threads, impossible to untangle. Alzheimer stared at them, transfixed. These tangles, he suspected, were just as destructive. They blocked the internal transport systems of the cells, starved them of nutrients, and, over time, killed them.

Here, at last, was a physical explanation. The confusion, the memory loss, the strange behaviour—it wasn't madness. It wasn't weakness. It was visible. Tangible. Embedded in the very structure of the brain.

Auguste's suffering had a cause. And now, perhaps, so did countless others.

It was the beginning of an answer—a discovery that could pave the way to understanding similar cases.

As the pieces of the puzzle began to fit together, Alzheimer felt something stir—part awe, part responsibility. He knew what he had found wasn't just rare. It was unprecedented. Auguste's illness had been more than a personal tragedy. It was a keyhole into a devastating disease that had gone unnamed, misunderstood, and unseen.

Now it had form.

Driven by a sense of urgency, he prepared to share his findings with the medical world. The setting was *the Tübingen Meeting of Southwest German Psychiatrists*—3rd November 1906. A routine professional gathering. But for Alzheimer, it carried the weight of revelation.

He took the podium with quiet determination.

Before a room of colleagues, he presented Auguste Deter's case in painstaking detail. Her early symptoms. The memory loss. The paranoia. Her incoherent speech. Her collapse into silence. Then, the breakthrough: the plaques and tangles embedded deep within her brain. A unique pathology. Something distinct from typical senile dementia. Something more insidious, and far earlier in onset.

He hoped—expected, perhaps—that his presentation would stir curiosity. That others might see what he saw: the possibility of a new diagnostic frontier. But as he spoke, a strange energy took hold of the room. Not excitement, but disinterest.

Eyes wandered. Notes were passed.

The next lecture—on compulsive masturbation—seemed to command far more anticipation.

When Alzheimer finished, the silence was suffocating. No questions. No discussion. Just a dismissive nod from the chairman:

"So then, respected colleague Alzheimer, I thank you for your remarks. Clearly, there is no desire for discussion."

The moment stung.

He had poured months into this research, carried the weight of a patient's life and death into a lecture hall, and been met with indifference. But he refused to let that silence have the last word.

Back in Munich, he set to work again—this time in ink.

In 1907, he published a short summary of his presentation. Later that year, he followed with a more comprehensive paper: *"A Characteristic Serious Disease of the Cerebral Cortex."* It laid out the clinical and pathological features of Auguste Deter's illness with clarity and precision. The plaques. The tangles. The cognitive decline. It was the first full description of what would, eventually, bear his name.

Three years later, the significance of his work found its champion.

In 1910, Emil Kraepelin included a new chapter in the eighth edition of his *Handbook of Psychiatry*, titled *"Senile and Presenile Dementias."* In it, he distinguished the condition described by Alzheimer from typical age-related decline.

And for the first time, the term *"Alzheimer's disease"* appeared in print.

A name. A category. A place in the medical literature.

It was a small entry, but it opened a door.

At the time, *dementia* was a broad and poorly understood label—its causes unclear, its course unpredictable. What Alzheimer had uncovered was something more specific: a disease marked by distinct changes in the brain, visible under the microscope. His patient, Auguste Deter, was unusually young, but those same abnormalities—amyloid plaques and neurofibrillary tangles—were soon found in older adults too.

Despite these milestones, Alzheimer's discovery did not spark immediate acclaim. There were no headlines. No sudden rush of recognition. But slowly—quietly—his work began to change the field.

By 1911, physicians across Europe and the United States were using his detailed case description to identify similar patients. Those whose symptoms had once been misattributed to hysteria, melancholy, or premature madness were now being seen through a different lens. His findings gave language to what had once been unspeakable. A framework. A diagnosis.

It was a shift not just in medicine, but in empathy.

In July 1912, Alzheimer's commitment was recognised in the most formal of ways. He was appointed full professor of Psychiatry and director of the *Neurological and Psychiatric Institute* at the *Silesian Friedrich-Wilhelm University* in Breslau. The decree was signed by Emperor Wilhelm II himself.

It should have been a moment of triumph.

But Alzheimer's health had already begun to falter.

During the journey to Breslau, he fell ill on the train. At first, it seemed like nothing—a seasonal fever, the wear and tear of travel. But the infection worsened, likely streptococcal in origin, and developed into rheumatic fever. It damaged his heart valves, setting off a slow, irreversible decline.

By early 1913, he was hospitalised in a private clinic. The man whose work had unravelled the mysteries of the brain was now battling the silent failures of his own body. His once-boundless energy was replaced by fatigue. His sharp mind dulled by illness.

Despite his determination to continue, his strength waned. Complications from heart and kidney failure followed. He lingered for nearly three years, drifting in and out of lucidity.

On 19th December 1915, Alois Alzheimer died in Breslau, Silesia (now present-day Wrocław, Poland). He was 51.

His death, like Auguste's before him, came quietly. A brilliant mind extinguished before its time.

And yet—he left behind something that refused to fade.

Though Alzheimer's life was cut short, his name would eventually become synonymous with one of the most devastating diseases of the modern era. His research, rooted in compassion, laid the groundwork for a century of discovery.

And at the heart of it all was Auguste Deter.

Her name is not as well-known as his, but she is everywhere in the story he helped tell.

Together, they serve as a reminder of the importance of understanding the fight against the loss of memory and identity.

What is Alzheimer's disease?

Alzheimer's disease is a progressive brain disorder that gradually erodes memory, thinking skills, and the ability to perform even simple daily tasks. It is the most common cause of dementia and is marked by the build-up of abnormal protein deposits in the brain—amyloid plaques and neurofibrillary tangles. These changes disrupt communication between brain cells, eventually causing them to die.

The disease often begins with mild symptoms such as forgetfulness or difficulty finding words, progressing over time to severe memory loss, confusion, and an inability to care for oneself. Beyond its physical effects, *Alzheimer's* greatly impacts not only those diagnosed but also their loved ones and caregivers, who witness the gradual loss of identity and independence.

While there is no cure yet, ongoing research is advancing our understanding of the disease and bringing hope for better treatments in the future. Raising awareness and fostering compassion for those living with *Alzheimer's* are vital steps in supporting individuals and families affected by this devastating condition.

CHAPTER 10

The Showman

JEAN-MARTIN CHARCOT STOOD just behind the door of the amphitheatre, his pulse quickening with a familiar blend of anticipation and unease. Public speaking never came easily to him, not even after years of practice. He had spent countless hours rehearsing for this moment, perfecting each word until it felt like second nature. But today wasn't just another lecture. It was a performance—one with the potential to reshape the very way medicine understood the human mind.

Beyond the heavy wooden door, the low murmur of the waiting crowd drifted through—curious, expectant, electric. He inhaled deeply, steadying himself. The stakes were always high in these demonstrations.

Joseph Babinski, his trusted junior physician, approached with Marie Wittman at his side. Babinski, as ever, was composed. He gave a subtle nod. "We're ready," he said quietly.

Charcot's gaze shifted to Marie. Her pale skin, lightly freckled, caught the soft glow of the gas lamps, giving her an almost spectral quality. Her wide eyes darted nervously, though her posture remained stiff, braced for what was to come.

"Thank you," Charcot murmured, nodding in return. He squared his shoulders, his hand brushing briefly against the cold brass handle of the door. For a fleeting moment, he wondered what Marie was thinking—what it felt like to be at the centre of such scrutiny, her private suffering laid bare for public observation. Then he pushed the thought aside and stepped into the light.

The room fell silent.

Rows of medical students, seasoned professionals, and curious onlookers filled the seats, their faces a sea of intent gazes. Some leaned forward, notebooks poised to catch every word. Others sat back with arms crossed; scepticism etched into their expressions. Charcot met their eyes, drawing a quiet strength from the weight of their attention.

His voice was steady as he began.

"Hysteria," he said, "has long been misunderstood. But today, we will witness its manifestations—and explore its deeper mysteries."

He gestured towards Marie, who now stood quietly beside Babinski.

"This is Mademoiselle Wittman. She has kindly agreed to assist in our demonstration. What you are about to see is not artifice. It is an accurate representation of the phases of hysteria."

He began by recounting the stages Marie had exhibited earlier that day. "The first phase is marked by sudden spasms and muscular contractions," he explained. "The body jerks uncontrollably. The eyes roll upward. The breath becomes rapid and shallow."

As he spoke, he glanced towards her, recalling the sheer violence of her convulsions. The way her small frame had contorted, jerking with a force that seemed to defy anatomy.

"In the second phase," he continued, "we observe dramatic contortions. The body may arch into what we call *arc de cercle*— where only the head and the heels touch the ground."

Murmurs rippled through the audience. Charcot's tone remained clinical, but the horror of what he described was difficult to mask entirely.

"Finally," he said, "the patient transitions through extreme emotional states—ecstasy, despair—before collapsing into delirium or extreme exhaustion."

His words hung in the air, uneasy and fascinating.

Then Babinski took over, beginning the hypnosis.

At the sound of the gong, Marie's body stiffened. Her left fist clenched, as though caught in the grip of an unseen force. Babinski moved with care, manipulating her limbs into rigid, statue-like poses—each one deliberate, each more disquieting than the last. The audience leaned in. Gasps punctuated the silence as Marie's body twisted into impossible positions. The boundaries of her flesh seemed to dissolve under suggestion.

Charcot narrated, his voice calm, though his eyes flickered with unease whenever they landed on her.

The transition to lethargy was even more striking. Marie's body collapsed, limp in Babinski's arms, as though life had drained from her entirely. Some spectators shifted in their seats, unsure whether what they were witnessing was science or something far more disturbing.

Charcot stepped forward. "This," he said, "is a profound state of muscular relaxation. But with suggestion, the muscles can contract again."

Babinski gave a soft prompt. Marie's body tensed—her posture once more rigid, deliberate, inhumanly precise.

Then came the third stage of hypnosis: somnambulism. In this heightened state, Marie became startlingly responsive. When instructed, she flinched and cowered in fear, arms raised to ward off an invisible assailant. A moment later, she transformed, her back straightening as she barked commands like a military officer. Her shifts in emotion were fluid and uncanny—uncannily theatrical.

The amphitheatre erupted into applause. A curious blend of awe and discomfort passed through the room.

As the demonstration concluded, Charcot turned back to Marie. She was waking slowly, her eyes glazed, her expression blank. She remembered nothing. The outbursts, the contortions, the strange commands—all gone. Her mind, once again, a clean slate.

Charcot felt it then—that strange fusion of triumph and guilt. To the audience, it had been a spectacle. A marvel of modern medicine. But to him, it was also something else: a sobering glimpse into the fragility of the mind, and the suffering it could endure in silence.

The applause faded.

The demonstration had been a success. But Charcot knew the real work lay not in performance, but in understanding. And understanding hysteria—truly understanding it—would take far more than public displays.

The painting "A Clinical Lesson at the Salpêtrière" by Pierre Aristide André Brouillet depicts Charcot demonstrating hypnosis on a patient named Marie Wittmann, with Dr Joseph Babinski standing behind her for support.

Jean-Martin Charcot was born in 1825 in a modest Parisian neighbourhood, still feeling the tremors of the French Revolution. The scent of iron and sawdust lingered in the family's small coach-building workshop, where his father laboured long days to keep the household afloat. Yet within those tight quarters and humble rhythms, Charcot's father held onto a quietly radical dream: that one of his sons might rise beyond the workshop walls and enter the world of higher learning.

He made a solemn promise—whichever son proved himself brightest would be given the rare chance to pursue an education.

For Jean-Martin, the eldest, it was both a gift and a burden.

His brothers, far from resenting him, rallied behind his efforts. On biting winter nights, as frost gathered on the windows and the air turned brittle with cold, they would take turns heating a small iron ball, nestling it in a bucket of sand, and placing it by his desk to warm the drafty little study. They lingered in the doorway just long enough to whisper encouragement before retreating to their own freezing beds.

"Stay warm, Jean," they'd say, voices soft in the dark.

He stayed up long after they'd gone, bent over Latin declensions, anatomical sketches, and candle-flickered textbooks. Outside his window, Paris moved on without him. Inside, he worked in silence.

His perseverance paid off. In 1844, he began studying medicine at the *University of Paris*. Four years later, in 1848, he earned a coveted internship at the *Salpêtrière*— a sprawling institution he would one day affectionately call the *"grand asylum of human misery."*

Nothing could have prepared him for what he saw there.

Originally a 16th-century gunpowder storehouse, the *Salpêtrière* had been repurposed into an immense and crumbling haven for society's outcasts. Over five thousand patients lived within its walls— those deemed too ill, too poor, too unruly to be anywhere else. Its corridors were endless and echoing, its rooms choked with damp and despair. The air was thick with the mingled odours of illness, neglect, and unwashed bodies. Cries of anguish echoed from distant wards.

Charcot walked its labyrinthine halls, absorbing everything. He saw women twisted by arthritis, their limbs frozen in pain. The glassy eyes of the mentally ill stared through him as though he weren't there. The frail clutched worn blankets, their silence more haunting than their moans. It was overwhelming. Yet he didn't turn away.

He made a quiet vow to himself—not just to treat these patients, but to see them. To understand them. To bring order, respect, and science to a place that had known too little of any.

In those early years, his focus was on chronic diseases. The *Salpêtrière* offered no shortage: arthritis, tuberculosis, cirrhosis, kidney failure. The work was unglamorous. Repetitive. But it demanded rigour, and Charcot gave it his full attention.

One day, while navigating a rarely used corridor, something unfamiliar drew him in.

A man stood in a dimly lit corner of the ward, hunched over a curious contraption. The device gave off a faint electrical buzz. Wires snaked outward, ending in small metal discs. As he touched the contacts to a woman's limbs, her muscles jerked—small, precise contractions that danced beneath her skin.

Charcot paused, fascinated.

The man turned, noticing Charcot's curiosity, and smiled. He gestured for him to come closer.

"This," he said, lifting the device slightly, "is a window into the nervous system."

His name was Guillaume Duchenne.

There was a quiet magnetism to him—measured, gentle, intensely focused. As he demonstrated the use of his induction coil, isolating individual muscles with uncanny precision, something in Charcot sparked. Here, before him, was a way to watch the body speak. To hear what muscles and nerves were trying to say when language failed.

"Every contraction tells us something," Duchenne said, his tone calm but electric. "The body speaks—if we're willing to listen."

Charcot asked questions—dozens of them. Duchenne answered them all, pleased by the younger man's curiosity. That moment marked the beginning of a transformative friendship.

Duchenne's techniques were as methodical as they were ground-breaking. He wasn't simply observing disease—he was tracing it, teasing out its mechanics, mapping its trajectory. Charcot began to look at his patients with new eyes. Not just as individuals weighed down by chronic illness, but as walking puzzles—each muscle tremor, each change in gait, a clue waiting to be decoded.

That encounter reshaped the course of Charcot's life.

When he later completed his thesis—distinguishing gout from chronic rheumatoid arthritis—it was with a precision born of this new neurological lens. The work earned him accolades and appointments: Chief of Clinic in 1853, and by 1856, Physician to the Hospitals of Paris.

Yet even as his success grew, something deeper was pulling at him.

The diseases of the body were no longer enough. What Duchenne had shown him—that the nervous system could be observed, charted, understood—had opened a door that couldn't be closed.

He began returning to the wards not as a physician of joints and organs, but of nerves. He noted tremors, reflexes, hesitations in speech and movement. He began to suspect that the body's afflictions might often be echoes of the brain's unspoken messages.

Neurology—still in its infancy—was becoming his true calling.

As Charcot delved deeper into the study of the nervous system, the *Salpêtrière* revealed itself not as a place of despair but as a vast, living archive of pathology. Its immense population, long neglected, became for him an invaluable resource—a human laboratory where every symptom had a story, and every story pointed toward a deeper truth.

He walked the wards daily, moving through their chaotic hum with a keen and hungry gaze. Behind him trailed students, interns, and junior doctors, all trying to keep pace. They watched him examine patients with an almost forensic attention to detail. He grouped cases meticulously, noting similarities others might miss—slight tremors, irregular speech patterns, subtle asymmetries in reflex.

What others saw as random suffering, Charcot began to systematise.

Even in death, his scrutiny continued. In the autopsy room, his hands were steady, methodical. With each incision, he pursued the invisible trail of neurological disease, drawing connections between the brain, spinal cord, and the symptoms he had seen at the bedside. What emerged was a vision of medicine that joined the seen and the unseen—what the body expressed, and what lay buried beneath the surface.

By the early 1860s, his efforts began to transform the *Salpêtrière* itself. In 1860, he was appointed associate professor. Two years later, he was named senior physician. Under his leadership, the once-forgotten asylum began its unlikely metamorphosis into a centre of scientific innovation.

The old stone wards that had echoed with weeping now buzzed with a different energy—one of inquiry, precision, and curiosity. With support from charitable organisations like the *Vincent de Paul Foundation*, Charcot introduced new technologies and techniques. He installed ophthalmoscopes, microscopes, photographic equipment. He established a state-of-the-art pathology laboratory. Patients who had once been warehoused were now studied, documented, and—perhaps most importantly—understood.

He wasn't merely categorising illness. He was reframing what it meant to observe. To notice. To care.

Yet, for all the demands of his professional world, Charcot's private life offered a counterbalance. He married Madame Durvis, a wealthy widow, and together they built a stable and affectionate home. Their two children brought him quiet joy, and on evenings when he wasn't dissecting or teaching, he would retreat to his study to sketch, reflect, or simply be still. His children, curious and adoring, often peeked through the doorway to catch glimpses of their father at work—part physician, part artist.

In those years, Charcot's gaze turned increasingly toward the mysteries of motor degeneration. He began to notice a cluster of symptoms in certain patients: twitching muscles, stiffness, and eventually, a frightening decline in speech and swallowing. In the final stages, paralysis set in. Breathing failed.

These were not isolated cases.

In the post-mortem room, he examined their spinal cords with the same fastidious care he gave to the living. He traced the slow death of motor neurons—those delicate threads that carried the body's commands from brain to muscle. What he found was consistent. Measurable. Devastating.

He named the condition *primary amyotrophic lateral sclerosis.*

Today, it is known simply as ALS.

It was Charcot who gave the disease a name. A language. A structure in which to grieve.

And yet, just as his work reached new heights, the world tilted once more.

In 1870, war erupted between France and Prussia (French-Prussian War of 1870-1871). As Paris descended into chaos, Charcot sent his family to England for safety. He remained behind, refusing to abandon his patients. The *Salpêtrière* filled not with neurological curiosities but with fever, infection, and death. Typhoid and smallpox swept through the wards. Charcot rolled up his sleeves and got to work—not as a researcher, but as a doctor in the most urgent and human sense.

Medicine, for him, was never about prestige. It was a calling.

In 1872, Charcot reached a career milestone that would cement his place among the greats. He was appointed Professor of Pathological Anatomy at the *University of Paris*—a title that, to most, spoke of textbooks and post-mortem dissections. But for Charcot, it meant something more. It offered a platform, a bridge between the classroom and the clinic, between the ivory tower and the aching, breathing bodies of the *Salpêtrière.*

And it gave him the freedom to teach in his own way.

On Tuesdays and Fridays, the grand amphitheatre at the *Salpêtrière* came to life. Long before the doors opened, students jostled for seats. Physicians and philosophers rubbed shoulders with curious aristocrats and visiting intellectuals. The room buzzed—not just with academic interest, but with anticipation.

When Charcot entered, the noise quietened instantly.

He was not a showman in the theatrical sense. Stocky, with a clean-shaven face and eyes that missed nothing, he commanded the room not with flourish but with gravity. He didn't speak for effect. His words were crisp, precise, layered with insight. Each demonstration unfolded like a well-planned composition. The pacing, the phrasing, the logic—all deliberate.

He would call forward a patient, interview them in real time, and then imitate their neurological symptoms with uncanny accuracy: the foot dragging of hemiplegia, the hand tremor of Parkinsonism, the vacant stare of catatonia. His impersonations were not mockery—they were teaching tools, a kind of embodied language.

For those in the room, it was unforgettable.

He was among the first to use photography as a diagnostic and teaching aid, capturing the physical nuances of disease that words alone could not convey. Enlarged prints were projected beside patients for comparison, creating a visual lexicon for neurology. His combination of clinical precision, observational skill, and artistry made him unlike anyone else in medicine.

And yet, for all his composure, there was something endearingly human about him.

Behind the stern academic exterior, Charcot harboured a love of mischief, music, and laughter. His wife, Madame Charcot, brought warmth and grace to their home, and together they created a haven of balance—an antidote to the intensity of the hospital.

Thursday evenings were sacred. No medicine. No lectures. No illness. Just music and merriment. The house filled with friends, students, and family. Laughter echoed through the halls. Charcot adored clowns and comic routines, often finding himself at the heart of the evening's joy. His humour was dry, sometimes teasing, but never cruel.

Saint Martin's Day on 11th November was a particular delight in the Charcot household. After a late supper, the children, their friends, and even some of his students conspired in an annual prank—designed specifically to unsettle the famously punctual professor.

Once the house had settled and the lights were low, the so-called "savages," as Charcot affectionately called them, would burst into his bedroom in wild formation—howling, dancing, shouting mock war cries.

Charcot, barely hiding his amusement, would feign outrage:

"Ah, those savages there! They do not know enough to go! If they do not leave at once, I shall arrive at *Salpêtrière* before my usual hour tomorrow morning—and then we shall see!"

But the act never lasted long. With a theatrical huff, he'd toss his bed slippers at them, laughter overtaking his scolding. These were the moments few outside his inner circle ever witnessed—the brilliant physician reduced to tears of laughter, surrounded by love.

Tuesday evenings, in contrast, took on a more refined tone. The Charcot home became a Parisian salon, alive with conversation and creativity. Artists, writers, scientists, and politicians gathered to share ideas beneath the soft glow of oil lamps and the gentle strains of piano music. No medical talk was permitted. It was a rule Charcot upheld firmly. The mind, he believed, needed more than intellect. It needed beauty, culture, contradiction.

Even in private practice, where he saw royalty and the global elite, his character never dimmed. He was patient and indulgent with the poor, charging them little or nothing. With the wealthy, however, his sharp wit occasionally surfaced. He had no tolerance for hypochondria draped in furs. Yet, despite his jabs, patients returned. His brilliance was magnetic—his warmth, unexpected.

One evening, following a particularly spirited salon, Charcot received intriguing news. A young woman named Marie Wittman had been admitted to the *Salpêtrière*'s epilepsy ward earlier that day, on 6th May 1877. She was just eighteen. Her symptoms resembled hysteria—nothing new to Charcot—but there was something about the case that stirred his attention.

Charcot began reading through Marie's file. What he found was not merely a list of symptoms, but a life marked by violence, illness, and survival.

Marie had been born in April 1859, in Paris, into a family steeped in hardship. Her father, a Swiss carpenter, was volatile and brutal. On one occasion, he beat her with such force that she was thrown from a window. Her mother, a linen maid, suffered frequent nervous attacks, often collapsing when upset. The home was steeped in chaos and unpredictability.

Of their nine children, five died—most from seizures or epilepsy. Death wove through Marie's early years like a recurring motif, each loss a silent lesson in fragility.

At just 22 months old, she experienced violent seizures that left her deaf and mute. She remained locked in that silent world for years. It wasn't until the age of seven that she began to speak and hear again. These early traumas carved deep grooves into her mind and body. School was fragmented, her education interrupted by illness and instability. She struggled to keep pace with other children, often isolated by her deficits and the sense that her life was somehow already separate from theirs.

At the age of 12, she was apprenticed to a furrier. The work was brutal—handling pelts, repetitive labour, long hours. Not long after, she began to suffer from night attacks: tremors, loss of consciousness, terrifying symptoms she managed to conceal, for a time. During the day, she grew increasingly clumsy. The tremors gave her away. Her employer accused her of malingering.

Then came the final violation. The furrier attempted to rape her. Marie fled.

She found temporary shelter working alongside her mother in a laundry. But this brief respite ended with her mother's death. Alone once again, she returned—desperately—to the furrier's shop. Whether out of poverty, necessity, or lack of options, the decision exposed her to further abuse. Eventually, she escaped again, this time into the care of a family friend.

Her journey spiralled through a series of unstable environments. She worked as a hospital duty maid. She entered a relationship with a young man named Alphonse, who, in his effort to "cure" her attacks, began applying painful pressure to her lower abdomen—a crude and misguided treatment reflecting the limited medical knowledge of the time. He pressed hard against her ovaries during her seizures, believing it might halt the symptoms. It didn't.

In search of peace, Marie retreated briefly to the countryside. Then returned to Paris. She sought sanctuary at a convent on Rue du Cherche-Midi. For a time, it seemed promising. But the attacks followed her. When the nuns realised her condition was beyond their care, they dismissed her.

Again, she was adrift.

And yet—she persisted. Determined to find help, she sought work as a servant at the *Salpêtrière*. It was a strategic move. She knew that as a staff member, she might one day be admitted as a patient. That was her aim: not escape, but treatment. A name for what plagued her. Relief.

On 6th May 1877, Marie Wittman was officially admitted.

From the moment Charcot met her, he was intrigued. She was delicate, pale, seemingly brittle—and yet beneath her surface lay a deep current of distress. Within a week, her attacks began. They were violent and prolonged. Her body seized with uncanny force. She contorted. She wept. She moaned and thrashed. At times, her face reflected emotions that were almost theatrical—ecstasy, terror, exaltation.

Charcot believed he was witnessing something entirely new.

He coined the term *hystero-epilepsy* to describe it—a disorder that mimicked epilepsy but bore the hallmarks of hysteria. Marie became his exemplar.

And so, she entered the amphitheatre.

From the moment she entered, Marie Wittman was no longer simply a patient. She became a focal point—a symbol of something larger, stranger, and, in Charcot's eyes, profound.

Her symptoms were dramatic, even mesmerising. She would convulse with remarkable force, her body arched in impossible postures, then fall limp and lifeless in seconds. At times she appeared overtaken by invisible forces, her expression locked in terror, then ecstasy, then military command. Each phase was vivid, uncanny. Her seizures didn't just disrupt—they performed.

Charcot, watching closely, believed he had uncovered

a neurological language the world had yet to decipher. Hysteria, long dismissed as emotional excess or feminine frailty, now stood before him as a patterned, observable, and perhaps treatable phenomenon. He named the condition. He categorised its phases. He defined its stages. Marie became the embodiment of his theory—living, visible, unforgettable.

The *Salpêtrière's* amphitheatre filled to capacity every Tuesday and Friday.

No longer a chamber of private instruction, it became Paris's most unlikely stage. Not only physicians and students, but artists, journalists, sceptics, and high-society women filled the seats. They came to witness Charcot's demonstrations—to see Marie fall into hypnotic trance at the sound of a gong, to watch her body transform at the suggestion of Babinski's guiding hand.

To the audience, it was science as theatre. A performance of pathology.

Marie, in her own way, became famous. Known in the corridors and lecture halls as the *"Queen of the Hysterics,"* she drew fascination and unease in equal measure. Some viewed her with pity, others with suspicion. Yet her role offered her something few women of her circumstances had ever known: visibility, even a strange form of power. In the act of being studied, she was also being heard.

Word spread far beyond Paris.

Among the captivated was a young Sigmund Freud. He sat in Charcot's lectures, taking feverish notes, watching as the unconscious seemed to unfurl itself in a woman's contortions. For Freud, the *Salpêtrière* was not just a hospital—it was revelation. He saw in Charcot's work the possibility of decoding the mind through suggestion, trauma, and memory. These early lectures would shape the very foundations of psychoanalysis.

In later years, Freud would name his first son Jean-Martin, in homage to the man who had opened his eyes.

Charcot, meanwhile, basked in acclaim. For the first time, hysteria had legitimacy—not just as a clinical curiosity, but as a diagnosable disorder. The ancient fear of wandering wombs and female madness had been replaced with something observable, classifiable, treatable.

His lectures cemented his place as a master of both neurology and narrative. Hysteria had found its champion.

But as the theatre grew louder, not everyone was convinced.

Behind the applause, a quieter voice was beginning to question the show.

Among the many who stood behind Charcot in the amphitheatre—who watched, learned, and assisted—was his most brilliant and loyal student: Joseph Babinski. Thoughtful, incisive, and fiercely respectful, Babinski had studied under Charcot for years. He respected him immensely, but that respect never clouded his judgment.

As Marie and others performed their hysterical symptoms before the crowd, Babinski began to notice something troubling.

Patterns.

Not the neurological kind Charcot had catalogued so carefully—but patterns of behaviour, of mimicry, of environmental response. Many of the women who exhibited seizures, contortions, and emotional transformations did so only under observation. When left alone, they were often symptom-free. Their most dramatic attacks occurred only in front of doctors. Or an audience.

Babinski also observed that the so-called *hystero-epileptics* often mirrored the epileptic patients they lived alongside. Their seizures became theatrical echoes—imitations shaped by exposure, not illness. It was, in its own way, a form of unconscious performance, crafted by expectation.

He began to suspect the unthinkable.

That the symptoms might not be a new neurological disease at all—but rather, a product of suggestion.

The idea was radical. Dangerous. And greatly personal. It called into question not just Charcot's theory, but the very environment he had created.

At first, Charcot resisted.

He was a scientist, but he was also human. For years he had built his reputation on the belief that hysteria could be studied like any other disease—that it had structure, stages, substance. To entertain the notion that it was being shaped by the setting, the observers— even by himself—was almost unbearable.

And yet, Babinski's logic was impeccable.

Quietly, carefully, Charcot began to look again.

He saw it now, where before he hadn't. The hypersuggestibility of certain patients. The way their symptoms aligned with what was expected of them. The unconscious cues they seemed to pick up from doctors, from demonstrations, from the weight of attention. The more sympathetic the staff, the more elaborate the fits.

It was a humbling revelation.

The amphitheatre, he now realised, was not just a stage—it was a crucible. It had helped *create* the very hysteria he sought to diagnose.

For all his intelligence and compassion, Charcot had underestimated one crucial force: the power of context.

In response, he made a quiet but significant change.

He and Babinski began transferring patients out of the specialised hysteria wards and into general medical ones. They reduced the attention. Removed the audience. Allowed the quiet to do its work.

And something extraordinary happened.

The seizures began to vanish.

Not always, not instantly—but in enough cases to confirm

Babinski's suspicion. Without the pressure of performance, without the presence of expectation, many patients calmed. Their symptoms eased. The theatre dimmed. And in its place, Charcot saw something deeper—a more complex, layered understanding of the mind's entanglement with the body.

He had not uncovered a new disease. But he had uncovered a truth.

That the body, especially under duress, will speak in whatever language it's taught to use. That patients, desperate to be seen, sometimes reflect the gaze that watches them. That even the most scientific environment can become, unknowingly, a mirror.

For Charcot, it was a moment of reckoning. And of grace.

He did not dismiss Babinski. He honoured him. He recognised in his student the same rigorous curiosity he had once felt standing in the wards with Duchenne.

The story of hysteria had changed. And Charcot, to his credit, changed with it.

Jean-Martin Charcot's fame reached extraordinary heights.

By the late 1880s, his name had become a fixture in international medical circles—and far beyond them. Letters arrived daily, from every corner of Europe and the Americas, from desperate patients, royal households, and rising stars of science. The crowd outside his private practice often stretched down the street. On the busiest days, Charcot would retreat not to a parlour or study, but to the bathroom—just to eat a quiet lunch in peace.

He had become a physician of emperors.

Among the many notable visitors was Grand Duke Vladimir Alexandrovich of Russia. Another was Emperor Pedro II of Brazil— an erudite monarch, fluent in multiple languages, who had followed

Charcot's work from afar and now came to see the master in person. Their consultations went beyond the clinical. In Charcot's elegant residence on Boulevard Saint-Germain, in the drawing room, amid books, art, and music, they spoke of science, politics, philosophy—and animals.

The emperor, a lover of natural history, was particularly struck by Charcot's gentle refusal to engage in animal experimentation. At a time when vivisection was commonplace in research laboratories, Charcot stood apart. He would not allow it in his hospital. A sign hung proudly on his office door:

"You will find no dog laboratory here."

He despised hunting. Called it barbaric. When asked about it, he would wave a hand dismissively, as if brushing away something foul.

Pedro II, moved by this rare compassion, found the perfect gift.

He returned some weeks later with a small, black-capped capuchin monkey—spirited, bright-eyed, and full of mischief. Her name was Rosalie.

She was an instant sensation.

Rosalie joined the Charcot household not as a pet, but as a member of the family. At mealtimes, she sat in an infant's chair at the table, daintily stealing nuts or slices of banana from Charcot's plate when she thought no one was watching. She darted through rooms, leaping from furniture with joyful abandon. Her antics became a nightly delight. Charcot, exhausted by the demands of his day, would burst into laughter at her games—his serious face softening, eyes twinkling.

She was, perhaps, the most unlikely medicine he had ever prescribed for himself.

Charcot's fame and success afforded him a lifestyle of comfort and cultural enrichment. Despite his demanding schedule, he found time to indulge in the arts, elegant soirées, and expeditions that allowed him to explore the world beyond the confines of his work.

Yet, as his career continued to soar, his health began to decline. His love of cigars and sedentary habits took a toll, and by his sixties, he was suffering from a heart condition, angina. The condition worsened over time, and his wife, ever attentive to his well-being, urged him to slow down.

But Charcot, driven by a passion for discovery, found it difficult to rest. In the summer of 1893, against his wife's pleas, he embarked on a cultural expedition to Vézelay with a group of his pupils. The journey was both invigorating and taxing.

Tragically, on 16th August 1893, Charcot suffered a severe fluid buildup in his lungs due to heart failure. In a modest guesthouse by the serene Lake des Settons, surrounded by his students, the great neurologist took his final breath at three o'clock in the morning. He was 67 years old.

The news travelled swiftly, a tremor through the medical world and beyond. Tributes poured in—from universities, monarchs, colleagues, and patients. His absence was felt like the sudden extinguishing of a great flame.

But the light he had kindled endured.

Over thirteen medical eponyms still bear his name. His clinical precision, his artistry, his teaching style, his diagnostic precision— these became cornerstones of modern neurology.

Charcot's son, Jean-Baptiste Charcot, chose to honour his father not by following in his footsteps but by carving his own path as a maritime explorer. Jean-Baptiste believed that success in a different field would best celebrate his father's spirit of exploration and discovery. In a gesture of respect, he named an island in Antarctica "*Charcot Island*," ensuring that the name Charcot would be forever linked to exploring uncharted territories; both in the human mind and the farthest reaches of the earth.

What was hysteria?

Hysteria was a term historically used to describe a broad range of psychological and physical symptoms, which were often misunderstood and disproportionately attributed to women. These symptoms included fainting, convulsions, paralysis, dramatic emotional outbursts, and other unexplained physical manifestations. At the time, hysteria was thought to stem from emotional disturbances or trauma, but it lacked a clear medical explanation, leaving patients stigmatised and often misunderstood.

Today, the term hysteria is no longer used in medicine. What was once broadly categorised under hysteria is now better understood as part of specific conditions, such *as Functional Neurological Disorder* (FND) and *Psychogenic Non-Epileptic Seizures* (PNES). FND is a neurological condition where patients experience real and often debilitating symptoms, such as movement difficulties, sensory changes, or weakness, without any detectable structural damage to the nervous system. PNES, meanwhile, involves seizure-like episodes that resemble epilepsy but are not caused by abnormal electrical activity in the brain. Both conditions are believed to arise from complex interactions between the mind and body, often linked to psychological stress or past trauma.

Modern medicine recognises that these symptoms are not fabricated. They are genuine, distressing, and deserve compassionate care. Patients are treated using a multidisciplinary approach that may include psychological therapy, physical rehabilitation, and medical support tailored to their individual experiences and needs. This shift in understanding has moved away from the stigmatising label of "hysterical" and toward a more empathetic and science-based approach to care.

CHAPTER 11

Storybook Syndromes

SO FAR, WE have explored how many conditions, manoeuvres, and scoring systems bear the names of notable figures from medical history. But did you know that some eponyms are drawn not from real people, but from fictional characters? These curious conditions often mirror the traits or afflictions of well-known literary figures, creating a compelling intersection between medicine and literature. In this chapter, we'll delve into several of these fascinating syndromes—examining their literary origins, the lived experiences of those affected, and the underlying pathologies that bind story to symptom.

Alice in Wonderland Syndrome

On a quiet Monday morning in the classroom, a nine-year-old boy sat at his desk, attentively watching his teacher write on the blackboard. Chalk tapped against the slate, mingling with the scratch of pencils as his classmates jotted notes into their copies. Everything seemed ordinary—until it wasn't.

At first, it was just the blackboard. It began to ripple, as though invisible waves were distorting its surface. The letters his teacher had carefully written shrank to the size of grains of rice, only to swell into warped, hulking shapes. The scene felt like a waking dream. He blinked rapidly, trying to force the board back into focus, but now the colours began to fade—the vibrant blue chalk, his classmates' clothes, even the teacher's dress. The world around him drained into lifeless monochrome.

Terrified, he glanced at the window beside him, hoping the glass might anchor him back to reality. Instead, he was met with a jarring reflection. His head, once familiar, had shrunk grotesquely small. Panic rose in his throat as he looked down. His left arm appeared impossibly long, broomstick thin. His hand, once slight and boyish, now swelled to the size of a balloon, comically distorted against the tiny desk.

Familiar sounds twisted into something monstrous. A chair scraped against the floor with the roar of thunder. Every whisper felt shouted directly into his ears. He pressed his hands to his temples, now pounding with a violent, rhythmic pain, and murmured, "Where am I?"

The classroom—his safe, predictable world—had become alien. Unfamiliar. Frightening.

Though the episode lasted no more than 15 minutes, it felt like an eternity. For weeks, the surreal nightmare replayed again and again, each time leaving him more frightened and exhausted. Twice a week, without warning, the bizarre sensations would return. His parents, initially dismissing his descriptions as overactive imagination, soon realised something was wrong. Their once vibrant and energetic son had become anxious, struggling to articulate the strange and other-worldly sensations he was experiencing. Desperate for answers, they sought the help of a neurologist.

This case is a classic example of *Alice in Wonderland Syndrome* (AIWS), a rare and fascinating neurological condition. AIWS is characterised by perceptual distortions and bizarre changes in body image, often making the sufferer feel they are shrinking or growing—just like Alice in Lewis Carroll's famous novel.

The term was coined in 1955 by British psychiatrist Dr John Todd to describe a cluster of symptoms *"intimately associated with migraine and epilepsy, although not confined to these disorders."* Symptoms may include *derealisation* (where the world seems unreal), *depersonalisation* (where the person seems unreal), illusory changes in the size, distance, or position of stationary objects, feelings of levitation, and altered perception of time.

Lewis Carroll—real name Charles Lutwidge Dodgson—famously described such experiences in *Alice's Adventures in Wonderland*. Carroll is thought to have suffered from migraines himself, a theory supported by his diary entries. In 1885, for example, he described *"that odd optical affection of seeing moving fortifications, followed by a headache."* These migraines, marked by vivid visual disturbances, may have inspired the surreal and disorienting imagery in his books.

Carroll's depictions of Alice's experiences resonate deeply with the symptoms of AIWS. When Alice falls down the rabbit hole, he writes:

> *"Either the well was very deep, or she fell very slowly, for she had plenty of time as she went down to look about her and to wonder what was going to happen next."*

This line captures the distorted sense of time and space typical of derealisation. Later, when Alice drinks from a mysterious bottle labelled "DRINK ME" and shrinks dramatically, Carroll writes:

> *"She was now only ten inches high, and her face brightened up at the thought that she was now the right size for going through the little door into that lovely garden."*

Moments later, she eats a cake and grows uncontrollably, disoriented by her enormous limbs:

> *"Curiouser and curiouser!" cried Alice (she was so much surprised, that for the moment she quite forgot how to speak good English); "now I'm opening out like the largest telescope that ever was!"*

These shifts in size and scale aren't just whimsical—they closely mirror the core symptoms of AIWS. Patients may see objects swell to overwhelming proportions (*macropsia*) or shrink to ant-like size (*micropsia*). Others describe looking at their own limbs and seeing them grotesquely distorted, or feeling their bodies expand, float, or dissolve altogether. The visual world becomes untrustworthy— warped by a mind that can no longer interpret size, shape, or distance correctly.

The earliest medical descriptions of such symptoms came from Caro Lippman in 1952, who documented patients experiencing striking sensations of height or shrinkage during migraine attacks. It was Dr Todd who gathered these disparate symptoms under one literary umbrella, naming the syndrome after Carroll's imaginative heroine and extending its definition to include visual and auditory distortions, altered time perception, and episodes of depersonalisation.

For those living with AIWS, Alice's story offers more than just a namesake. It becomes a symbolic vocabulary—a poetic means to express the inexpressible. For children like the boy in our story, Wonderland feels perilously close to home, a reminder of how fragile and malleable perception can become.

The cause of AIWS lies in the brain's perceptual circuitry, particularly the temporo-parietal-occipital junction—a region responsible for integrating sensory and spatial information. When this network is disrupted—by migraines, epileptic activity, viral infections (such as Epstein–Barr), or other neurological insults—the mind's internal map of the world collapses in on itself.

Beyond visual distortion, AIWS can warp time, dissolve identity, or uncouple a person's sense of self from their environment. These episodes, though often brief, are extremely unsettling. For a child, they can turn the mundane into the menacing, transforming a classroom into a hall of mirrors.

Alice's descent into Wonderland was fantastical. For those with AIWS, the descent is real and they're not chasing rabbits, but their own lost sense of reality.

Munchausen Syndrome

In a quiet suburban neighbourhood, a 28-year-old housewife named Emily found herself once again in hospital—this time with tender, reddish lesions scattered across her body. They had appeared suddenly two days earlier, causing severe pain, but no fever or other symptoms. It wasn't the first time Emily had experienced such an episode. Over the past nine months, she had been admitted to various hospitals on four separate occasions, each visit uncannily similar to the last.

Her medical history puzzled the doctors. There was no history of prolonged sun exposure, no new medications, no apparent allergies—nothing to explain the appearance of the lesions. Each time she was hospitalised, Emily would leave before completing her treatment. The pattern repeated: reddish-brown spots, tender to the touch, scattered across her thighs, legs, chest, abdomen, face, and—curiously—sparing her back.

On examination, the lesions appeared uniform in size and shape. Strangely, when scrubbed with alcohol, some of them faded in colour. Despite exhaustive testing—including blood work, urine analysis, and skin biopsies—no definitive medical cause could be identified. The medical team decided to administer placebos, without informing Emily or her family. Within a week, the lesions began to fade.

A closer look into Emily's psychological state revealed no overt hallucinations or perceptual disturbances. There were no obvious external motives—no avoidance of responsibilities, no financial incentives, no litigation. Yet both Emily and her husband gave inconsistent accounts of events, further muddying the waters. Notably, each episode coincided with her husband being away for work.

For Emily, each hospital visit wasn't just about the lesions. It hinted at something deeper—a silent plea for attention that no blood test or biopsy could uncover. While the doctors searched for a diagnosis, Emily seemed to be searching for something else: perhaps validation, perhaps compassion that went beyond the physical. Her emotional distress was evident, even if unspoken, adding a profound layer of complexity to her case.

After thorough evaluation, the team concluded that Emily's condition was not dermatological, but psychological. The uniformity of the lesions and her behaviour pointed towards self-inflicted fabrication. It was suspected she had used substances like makeup, crystallised sugar, or dye to mimic skin disease—materials that would discolour the skin but could be easily wiped away with alcohol. The placement of the lesions—easily reached areas of her body, sparing the back—supported this theory. Under this new understanding, her lesions resolved completely within two weeks, and she was referred for psychiatric care. Her final diagnosis: *Munchausen syndrome.*

Munchausen syndrome, first described in 1951 by Dr Richard Asher, refers to individuals who deliberately produce signs and symptoms of illness in order to seek medical attention through deceptive means. In 1977, the related term *Munchausen syndrome by proxy* was introduced by paediatrician Roy Meadow, describing cases in which caregivers—most often mothers—fabricate or induce illness in their children to attract sympathy or admiration. This form of the syndrome may also extend to the elderly or other dependent adults under a caregiver's control and is now classified under the broader framework of abuse against vulnerable individuals.

The inspiration for the syndrome's name comes from the fictional character Baron Munchausen, loosely based on the real-life German nobleman Hieronymus Karl Friedrich von Münchhausen. The fictional Baron became famous for his wildly exaggerated and fantastical tales, immortalised in the late 18th century by Rudolf Erich Raspe in Baron Munchausen's *Narrative of his Marvellous Travels and Campaigns in Russia*, published in 1785.

In Raspe's telling, Baron Munchausen is a boastful and eccentric adventurer who recounts a series of impossible feats with absolute conviction. His stories, filled with impossible feats and absurd situations, include trips to the moon, escaping from a whale's belly and riding a cannonball. For example, in one tale, the Baron saves himself from drowning by pulling himself out of a swamp by his own hair. In another tale, he harnesses a team of wolves to his carriage and speeds across the Russian steppe. These absurd, imaginative tales often parodied the excesses of contemporary travel literature and the gullibility of his audience.

Much like the Baron, individuals with *Munchausen syndrome* spin elaborate, implausible narratives—though theirs are not amusing diversions, but emotionally layered, often tragic expressions of psychological distress. Their symptoms may be self-inflicted or entirely fabricated but are presented with conviction and emotional urgency. The resulting deception is not rooted in malingering or external reward, but in a compulsive need to inhabit the role of the patient, often in a desperate attempt to receive care, attention, or validation.

A hallmark feature of the syndrome is *pseudologia fantastica*, a term used to describe compulsive, elaborate lying. These fabrications are often internally consistent and plausible enough to mislead even experienced clinicians. Some patients possess a striking familiarity with medical jargon, tests, and procedures—knowledge gained through repeated hospital visits or, in some cases, professional training within healthcare settings. They may willingly undergo invasive investigations, seem oddly comfortable with hospital routines, and disappear once suspicions begin to surface.

The psychological origins of *Munchausen syndrome* are multifaceted. Many patients have histories of childhood trauma, emotional neglect, or insecure attachments. Others may display masochistic tendencies or suffer from fragmented identities and an intense yearning for care and containment. In some cases, these behaviours are thought to reflect an unconscious attempt to externalise internal suffering—a symbolic cry for help dressed in the trappings of medical illness.

Treating individuals with *Munchausen syndrome* is especially challenging. Clinicians must navigate a careful path between uncovering deception and maintaining empathy. Strategies such as the use of placebos, as employed in Emily's case, may offer diagnostic clarity, but they carry ethical risks. The therapeutic alliance can be fragile, and confrontation—if handled without tact—may drive the patient away entirely.

Yet recognising the condition is crucial, not to discredit the patient, but to redirect focus toward their underlying needs. It shifts the question from *What is the illness?* to *Why is this person unwell?* — a subtle but essential reframing that allows for psychological engagement rather than endless medical investigation.

In the end, Emily's story is not just about skin lesions. The outward signs were a mask for something more fundamental and far less visible: the aching need to be noticed, understood, and cared for. For those with *Munchausen syndrome*, the symptoms are rarely the disease itself—but rather the language through which their suffering is expressed.

Othello Syndrome

Linda's story begins on a warm summer evening. She sat alone in her small kitchen, a cup of untouched tea cooling on the table. At 67 years old, Linda had already weathered more than most: multiple strokes, a haunting childhood trauma involving accidental lye poisoning, and a long battle with poorly controlled diabetes. But now, she faced a new torment—one that unravelled the fabric of her marriage from the inside out.

Her husband, John, had been her steadfast companion for over four decades. They had raised a daughter, survived financial hardship, and celebrated life's quiet victories together. But recently, Linda had become consumed by a creeping suspicion that John was being unfaithful. It began as a vague discomfort, a fleeting doubt she tried to dismiss. But the feeling grew, taking root in her thoughts, fed by John's decreasing interest in intimacy and his occasional use of pornography.

Convinced of his betrayal, Linda installed cameras throughout the house. She spent hours hunched over the footage, scouring the images for clues. The shadows and blurry reflections became women in her mind—silent intruders she was sure John had invited in. She showed the clips to her daughter, pointing at empty doorways and indistinct shapes, her voice trembling with certainty. Her daughter, bewildered, could see nothing. But Linda's belief only deepened.

She cut back her work hours to monitor the house more closely. Sleep became scarce. Her eyes, once lively, turned glassy with exhaustion. Every minute of her day circled back to the footage, to the quiet devastation of not being believed.

John, patient and soft-spoken, tried to reassure her. He answered every question, denied every accusation, but it wasn't enough. Linda's mind looped endlessly through doubt and conviction. Their home, once calm and filled with easy laughter, became a battleground of suspicion and silence. Linda cried often, retreated into herself, and sank into long spells of despair that John felt powerless to comfort.

During a hospital admission following a stroke, Linda's psychiatric evaluation revealed the scope of her fixation. She spoke clearly, calmly even, describing her suspicions with a flat affect and moments of quiet tearfulness. Her speech was coherent, her thoughts linear—but every road led back to John's imagined infidelity. She denied hallucinations and suicidal thoughts, but her belief remained unshakeable.

After discharge, the delusions persisted. Their marriage, already fraying, began to unravel under the strain. Family gatherings grew tense. Linda withdrew from friends, ashamed of the accusations she couldn't stop making. Despite her emotional deterioration, she resisted psychiatric treatment. Her pride and a deep mistrust of medication created yet another barrier between herself and the help she needed.

For John, the emotional toll was devastating. He walked on eggshells, afraid that any stray comment or misplaced object might spark another spiral. And yet, he remained—anchored by love, by hope, and by the memory of the woman he had known before the storm.

Linda's experience is a striking example of *Othello syndrome*, a psychotic disorder characterised by delusions of infidelity and jealousy. First coined by Todd and Dewhurst in 1955, the condition often arises in the context of medical, psychiatric, or neurological disorders—particularly in patients with dementia. It manifests as a fixed false belief that one's spouse is engaged in extramarital affairs, leading to emotional devastation and, at times, violence. Clinically, it's associated with several neurological disorders, including neurodegenerative disease and stroke. Coexistent delusions and hallucinations are frequent, and violent behaviour in affected patients, especially those with dementia, is well documented.

The eponym originates from Shakespeare's tragedy *Othello, the Moor of Venice*. In the play, Othello—a Moorish general in the Venetian army—is manipulated by his deceitful ensign, Iago. Through lies and forged evidence, Iago convinces Othello that his wife, Desdemona, has been unfaithful. Consumed by jealousy, Othello murders her in a fit of rage, only to discover the truth too late. Overwhelmed by guilt, he takes his own life.

Shakespeare's portrayal of jealousy as the "green-eyed monster" captures its corrosive power—how it mocks and devours the very person who harbours it. As Iago warns:

> *"O, beware, my lord, of jealousy;*
> *It is the green-eyed monster which doth mock*
> *The meat it feeds on."*

While Othello's story is set within a marriage, the syndrome can emerge in any intimate partnership. Patients often gather "evidence" from benign or unrelated events—overheard conversations, misplaced objects, shadows in a photograph—and weave them into a narrative of betrayal. This belief persists despite reassurances, denials, or the absence of any proof.

Othello syndrome occurs across both functional and organic psychoses. It's seen in paranoid schizophrenia, mood disorders with psychotic features, and a range of neurological conditions including stroke, *Parkinson's disease*, brain trauma, tumours, encephalitis, and multiple sclerosis. Delusions of infidelity may arise in the setting of frontal or temporal lobe dysfunction, particularly following strokes.

Linda's case mirrors the classical features of the syndrome. Her life—once stable and rooted in the deep trust of a long marriage—became dominated by the obsessive conviction that John was being unfaithful. Her decision to reduce her work hours, the sleepless nights, and the endless review of security footage all illustrate the consuming nature of the delusion. The belief had become its own self-reinforcing reality.

Her background added another layer of complexity. With multiple strokes and poorly controlled diabetes, Linda had already experienced significant neurological insult. These changes may have contributed to the onset of her delusions, disrupting circuits involved in reasoning, emotional regulation, and reality testing.

But Linda's story is not unique. Auguste Deter—discussed earlier in this book—also displayed signs of delusional jealousy toward her husband, Carl, as part of her descent into dementia. Her case, which helped define what would later be recognised as *Alzheimer's disease*, reminds us that jealousy and paranoia often emerge as part of cognitive decline. Like Linda, Auguste's accusations reflected not a loss of love, but a loss of the brain's ability to correctly interpret the world around her.

Othello syndrome is a window into that devastating collision between love, memory, and mistrust. The delusion fractures relationships at their core, turning the familiar into something menacing. For partners like John, the emotional toll is immense. They often feel powerless, living in a state of constant defence, where every word or gesture may be misinterpreted. It becomes a kind of mourning: grieving the partner who still stands before them, but who no longer sees them clearly.

Pickwickian Syndrome

The poker room buzzed with low conversation and a haze of cigarette smoke, the tension of the game wrapping tightly around its players. At the centre of the table sat Richard, 51-year-old business executive whose life had quietly spiralled into a vortex of stress and overeating. His thoughts flickered like the cards in his hand.

"Full house," he mused, glancing down at three aces and two kings. "This should be my night."

But as he leaned forward, the harsh truth pressed against him—literally. His bloated stomach nudged the edge of the table, a stubborn reminder of the weight he'd gained over the past year. Stress had driven him to seek comfort in food, and the more he ate, the more the weight piled on. The heavier he became, the more his symptoms worsened.

The vibrant colours of the poker chips began to blur. His eyelids grew heavy. The familiar hum of the room faded into a muffled drone. Before he knew it, Richard was asleep, his full house forgotten as his head drooped forward and he began to snore softly. The other players exchanged puzzled glances, their surprise at his sudden nap mingling with the competitive spirit of the game.

Days later, Richard found himself in a far less forgiving setting: the sterile, over lit corridors of the *Peter Bent Brigham Hospital*. He could no longer ignore the signs. He was falling asleep during meetings, sometimes even while standing. The blurred line between dreams and waking, the short, shallow breaths, the swelling in his ankles, the fainting spells—all of it had culminated in that moment at the poker table.

Richard sat there as the doctor examined him, struggling to keep his eyes open. His weight had ballooned to 120kg, starkly contrasting the 90kg he'd maintained for years. At 5 feet 5 inches tall, the extra weight made every movement a struggle. His face was ruddy, and the bluish tinge around his cheeks and nails was hard to miss. He tried to focus on the doctor's words, but the overwhelming urge to sleep pulled at him.

"Richard, your condition is severe," the doctor said gently. "Your weight, combined with the fatigue and sleepiness, points to something deeper."

Richard nodded, though he barely registered the words. His eyes fluttered closed, and he began to snore softly. The doctor sighed and made a note on his chart.

Throughout the consultation, Richard's breathing remained erratic—rapid and shallow, broken by brief silences and twitching movements as he drifted in and out of sleep. Once a confident and articulate executive, he now found himself increasingly withdrawn, embarrassed by his inability to remain awake even during conversations. When he was alert, he spoke clearly and cogently. But as soon as he relaxed, the cycle would begin again—eyes closing, body slumping, awareness fading. He was trapped in a body that refused to stay awake.

Richard's story mirrors the struggles faced by many individuals with *Pickwickian syndrome*, more formally known as *Obesity Hypoventilation Syndrome* (OHS). This condition arises when a person's breathing becomes too shallow to maintain normal levels of oxygen and carbon dioxide in the blood. It is primarily caused by severe obesity, which places immense strain on the respiratory system. Unlike many other breathing disorders, OHS is directly linked to the physical burden excess weight places on the chest and lungs.

The name *"Pickwickian syndrome"* originates from Charles Dickens' *The Posthumous Papers of the Pickwick Club.* In it, Mr Wardle's servant boy, Joe—an obese and perpetually drowsy figure— frequently falls asleep without warning. Dickens captures him vividly in an early scene:

> *"Damn that boy,"* said the old gentleman, *"he's gone to sleep again."*
>
> *"Very extraordinary boy, that,"* said Mr Pickwick, *"does he always sleep in this way?"*
>
> *"Sleep!"* said the old gentleman, *"he's always asleep. Goes on errands fast asleep, and snores as he waits at the table."*
>
> *"How very odd!"* said Mr Pickwick.
>
> *"Ah! Odd indeed,"* returned the old gentleman; *"I'm proud of that boy—wouldn't part with him on any account— damme, he's a natural curiosity!"*

Dickens' portrayal of Joe was meant to be comedic, but his symptoms—daytime sleepiness, unrefreshing sleep, and obesity—are recognised as classic features of *Pickwickian syndrome.*

The pathophysiology of the condition centres on the mechanical impact of obesity. Excess fat around the chest and abdomen restricts the movement of the diaphragm and lungs, reducing the depth and efficiency of each breath. As a result, the lungs struggle to draw in enough oxygen or expel carbon dioxide effectively. At night, when muscle tone naturally decreases during sleep, these effects are compounded. The airway may partially or completely collapse, leading to *obstructive sleep apnoea*—a condition frequently associated with *Pickwickian syndrome*. Over time, carbon dioxide builds up in the bloodstream, and the body slips into a state of chronic respiratory acidosis.

In simpler terms, the very mechanics of breathing begin to fail under the weight of the body itself. The result is a daily battle with exhaustion—not just a physical fatigue, but a neurological one, driven by the brain's increasing inability to regulate wakefulness under the stress of low oxygen and high carbon dioxide levels.

This isn't just physical fatigue—it's a constant battle to stay awake, even during important conversations or, in Richard's case, a poker game, as he finds himself irresistibly drawn into sleep at the most inopportune times, much like the character Joe in Dickens' classic tale.

Managing *Pickwickian syndrome* is complex. It involves weight reduction through dietary changes, physical activity, and sometimes bariatric surgery. Respiratory support, particularly at night, can be life changing. *Continuous Positive Airway Pressure* (CPAP) machines help maintain open airways during sleep, improving oxygenation and reducing the buildup of carbon dioxide.

Yet diagnosis is often delayed. Patients and clinicians alike may attribute symptoms to "just being overweight" or "overtired," overlooking the deeper physiological failure beneath. For Richard, the moment at the poker table wasn't just an embarrassment—it was a tipping point. A wake-up call, quite literally, that his health was collapsing in ways he had not understood.

Rapunzel Syndrome

Rachel was a vibrant 15-year-old, known for her spirited nature and a smile that seemed to light up any room. Yet behind those sparkling eyes, she harboured a secret struggle—one that would unravel in a way no one expected.

It began quietly. Rachel complained of intermittent stomach pain, the kind that came and went, sometimes accompanied by nausea and vomiting. Her appetite faded, and she often felt uncomfortably full after just a few bites. At first, her parents dismissed it as a passing stomach bug, but as the episodes grew more frequent, their concern deepened.

One evening, after another spell of pain left Rachel pale and exhausted, her parents decided enough was enough. At the doctor's office, a gentle examination revealed a firm, movable lump in her upper abdomen, roughly the size of a small cucumber. It was a curious finding—one that called for further investigation.

A CT scan of Rachel's abdomen revealed a large, well-defined mass within her stomach. The images showed an unusual swirl of densities, with pockets of trapped air—hardly the appearance of a typical growth. To get to the bottom of it, her doctor recommended an upper gastrointestinal endoscopy—a flexible camera threaded down the throat to peer inside the stomach. What they found stunned the medical team.

Rachel's stomach was almost entirely filled with a huge, tangled mass of hair. This hairball extended from her stomach into the first part of her small intestine. The sheer size of the mass made removal with standard tools impossible, so the team decided on surgery.

During the operation, the surgeons carefully opened Rachel's stomach and removed the enormous lump. It was tightly packed with hair, interwoven with undigested food, and gave off a foul odour—a grim testament to how long it had remained hidden. For a moment, the room was silent as everyone took in the scale of what they'd extracted. The hairball's size and shape told a story of years of quiet struggle.

In the aftermath, Rachel found the courage to confess what she had never shared aloud. For as long as she could remember, she'd felt an irresistible urge to pull out and eat her own hair. This behaviour,

known as trichophagia, had become a secret coping mechanism—
one she felt powerless to resist. Her parents listened; a blend of
sadness and compassion etched on their faces. Rachel's admission
was the first step towards understanding a condition that had, until
then, remained locked away in silence.

Following her surgery, Rachel underwent a comprehensive
psychological assessment. She was diagnosed with trichotillomania—a condition marked by the compulsive urge to pull out one's
own hair. Coupled with trichophagia, this behaviour had led to the
formation of a massive trichobezoar—a densely packed hairball that
had filled her stomach and begun extending into her intestines.
Treatment focused on helping Rachel develop healthier coping
mechanisms through cognitive behavioural therapy and emotional
support. Slowly, with guidance and the steady encouragement of
her family, Rachel began to regain control of her urges and take her
first steps toward recovery.

Rachel's condition is a striking example of *Rapunzel syndrome*, a rare
and unsettling disorder named after the long-haired maiden in
the Grimm Brothers' fairy tale. In the original story, Rapunzel—a
twelve-year-old princess—is locked away in a tower by an enchant-
ress. With no stairs or doors, the sorceress climbs up to her using
Rapunzel's long, golden hair. The tale, steeped in themes of isolation
and eventual escape, finds eerie resonance in Rachel's own experience: a physical entrapment, hidden and unspoken, that she too
needed to be freed from.

Rapunzel syndrome was first described in 1968 by Vaughan and
colleagues. It refers to a specific presentation of trichobezoar in
which the mass not only fills the stomach but also trails into the
small intestine, sometimes even further. The condition primarily

affects adolescent girls between the ages of 13 and 20, often those with underlying psychiatric disorders such as trichotillomania and trichophagia. The ingestion of hair—unlike other foreign bodies—results in a substance that the body cannot digest or pass naturally. Over time, the hair accumulates in the folds of the stomach, tangling into a dense, matted ball that conforms to the shape of the stomach itself.

As the trichobezoar grows, it begins to press against the stomach wall and obstruct the passage of food, leading to symptoms like abdominal pain, early satiety, weight loss, bloating, and vomiting. Left untreated, it can cause serious complications: gastrointestinal perforation, peritonitis, ulceration, and in rare cases, life-threatening obstruction. Surgical removal is often the only option for large bezoars, as in Rachel's case.

But the physical consequences, however dramatic, are only half the story. *Rapunzel syndrome* speaks just as powerfully to the hidden emotional landscape from which it arises. Trichophagia often develops in the context of unresolved trauma, anxiety, or a need for self-soothing. The act of pulling and eating hair becomes an unconscious ritual—a means of expressing distress that cannot be spoken. Rachel's story is not just a medical anomaly; it's a reminder of how the mind and body entwine in strange, sometimes dangerous ways.

Recovery requires more than just surgery. Without psychological support, the underlying behaviours may persist, leading to recurrence. But with the right care, healing is possible. Rachel's progress reflects that journey—not just the removal of a foreign mass, but the beginning of an emotional unburdening.

Rapunzel, in the original tale, eventually finds freedom not by cutting her hair, but by confronting the circumstances of her entrapment. Likewise, Rachel's liberation came not simply through surgical intervention, but through the courage to speak about what had bound her in silence.

CHAPTER 12

The Art and Anatomy of War

THE JANUARY WIND howled through Portsmouth's narrow streets, carrying with it the sharp tang of salt and a sense of unspoken dread. The port town, usually alive with the chatter of merchants and the clamour of dockworkers, was hushed, its lively spirit subdued by the looming threat of conflict. News of the British Army's retreat from Corunna, Spain, had swept through like wildfire, leaving behind a trail of fear and grim anticipation. The wounded would soon arrive, and the town braced itself for the storm of pain and chaos to follow.

It had been just over a week since the Battle of Corunna—a desperate clash on 16th January 1809, during the Peninsular War. British forces, battered and outnumbered, had fought Napoleon's army to protect their retreat to the sea. Victory, if it could be called that, came at a terrible cost. The Peninsular War formed part of the larger Napoleonic conflict: a bitter struggle in which British, Portuguese, and Spanish forces fought to liberate the Iberian Peninsula from French occupation. Now, Portsmouth would bear that cost—its quaysides transformed into makeshift hospitals for the broken men who had lived through it.

As the first ships docked beneath a bruised grey sky, the scene on the quay erupted into a kind of controlled chaos. Soldiers—gaunt and ghostlike—were carried ashore on stretchers cobbled together from whatever the crews had managed to find. The air hung thick with the mingled cries of the wounded, the bark of officers issuing orders, and the constant stomping of boots on wet planks. Blood stained the timbers, mixing with seawater in dark, briny streams that seemed to drain the very essence of life from the injured.

Amid the frenzy, one figure moved with calm purpose, his presence a quiet anchor in the turmoil. Charles Bell, a young surgeon, stepped onto the quay with measured strides. His tall frame and steady bearing made him stand out—not by appearance, for his coat had clearly seen better days, and his leather bag of surgical tools was worn at the seams—but by the look in his eyes. There was a quiet intensity there, a gaze that cut through the chaos as if he were seeing something deeper, something worth understanding.

Bell had volunteered to be here, though he was no military man. His reputation as an anatomist and surgeon had already begun to grow—whispered about in medical circles for his steady hands and brilliant mind. Yet there was no ceremony in his arrival. He set to work at once, boots splashing through shallow pools of blood and brine as he approached the first stretcher.

A young soldier lay before him, his face pale and twisted with pain. A musket ball had torn through his shoulder, splintering the bone and leaving an angry, jagged wound. Bell crouched beside him, his voice calm and reassuring as he spoke to the lad—though the soldier's glazed eyes showed little recognition. The surgeon's hands moved with practised precision, cutting away the blood-soaked fabric and cleaning the injury before beginning the grim task of removing the bullet.

Nearby, another stretcher arrived. This man's chest heaved in shallow, rattling gasps. Bell's gaze sharpened as he knelt beside him, fingers probing gently along the torn flesh, tracing the path of the shot. "Hold him steady," he said, low but firm, his tone leaving no room for delay. The wound was bad—very bad—but Bell moved with an almost balletic grace. The bullet came free, the bleeding was stanched, and the patient stabilised. Each movement was deft and assured, the product of years of study and curiosity.

But it wasn't just his skill with a scalpel that set Charles Bell apart. As the first wave of wounded eased and other surgeons paused to rest or tend to the living, Bell did something that drew quiet glances from those around him. He pulled a small sketchbook from his bag— its leather cover smoothed with years of use—and a pencil from his pocket. Sitting on an upturned crate, still streaked with blood and grime, he began to draw.

It was a strange sight: a man who had just spent hours knee-deep in carnage now sitting quietly, sketching with the focus of a man working on a masterpiece. His pencil moved swiftly, capturing the unflinching truth of what lay before him. The shattered bone from the musket wound, the torn muscle, the ragged trauma to the chest—each detail was committed to the page with a precision that seemed almost out of place amid the clamour.

"What's he doing?" one surgeon muttered to another, glancing at Bell.

"Sketching, by the look of it," came the reply, laced with confusion and a flicker of admiration. "Strange man."

But Bell paid no attention to the whispers. To him, the sketchbook was as necessary as the scalpel. He wasn't merely treating injuries—he was documenting them. Each line and shadow was a study in human fragility and resilience, a visual record of the war that words alone could never fully convey. His fascination with anatomy was unmistakable in every stroke, but this was more than science. This was witness. Preservation. A way to make sense of the suffering unfolding around him.

As the sun dipped low on the horizon, casting a burnt-orange glow over the harbour, Bell slipped the sketchbook back into his bag. His coat was stiff with dried blood, his hands rubbed raw from the day's work. He rose and looked across the makeshift hospital at the wounded men who lay scattered across it.

He had no way of knowing that the skills he honed here would become essential in the aftermath of the Battle of Waterloo.

As a child in Edinburgh, Charles Bell would spend hours with charcoal in hand, sketching the people and places around him with a quiet intensity that seemed unusual for his age. His mother, ever watchful of her children's gifts, noticed something unusual in her youngest son. Keen to nurture that spark, she arranged for him to study under David Allan, a respected Scottish painter. Under Allan's guidance, Bell's drawings began to evolve. What began as simple portraits of family members soon turned into intricate studies—of hands, muscles, and fleeting expressions. He lingered over details others might have skimmed past: the delicate sweep of a tendon, the fine crease of a brow mid-thought.

By the time he was a teenager, his fascination with the human form had deepened into something more enduring—something that would shape the whole of his life.

Charles's older brother, John Bell, was already a formidable figure in Edinburgh. A celebrated surgeon and anatomist, he was known for both his brilliance and his sharp tongue. Despite John's intensity, he saw something of himself in his younger brother. Charles began assisting him in the dissection room and in his writings. Amidst the clinical scent of preservative fluids and the cold gleam of steel instruments, Charles found his calling.

It was not merely observation. He contributed. He helped illustrate John's monumental four-volume work, *The Anatomy of the Human Body*, capturing in ink what words alone could not always convey. His drawings were not dry diagrams—they were studies of precision, each one infused with clarity and care. When Charles Bell later published his own illustrations in *A System of Dissections*, it was clear he was no longer standing in his brother's shadow.

Driven by a restless curiosity, Bell opened an anatomy museum in Edinburgh. The space was unlike anything most had seen—a gallery of the human body laid bare. Skeletons hung mid-air, organs floated in jars of preservative, their surfaces gleaming beneath flickering candlelight. His drawings adorned the walls, bridging the divide between science and art. Medical students came to study the body's inner scaffolding, their fingers tracing the inked annotations. Artists came too, sketching muscles and joints to refine their own portrayals of the human form.

To Bell, this museum was far more than a teaching space. It embodied his belief that art and medicine were not separate pursuits, but reflections of the same truth. He often lectured to rooms that held both doctors and painters, his voice carrying with conviction.

By the time Charles married Marion Shaw in 1811, his reputation had spread far beyond Scotland. The newlyweds settled in Soho Square, London, where Bell took over the prestigious anatomy school on Great Windmill Street. The city's vibrant core became the backdrop to his rising influence as a teacher, surgeon, and writer. Students came from across Europe to learn from the man whose scalpel and pencil were wielded with equal skill.

It was there, in London, that Bell made one of his most groundbreaking discoveries. In a cramped laboratory, under the dim flicker of oil lamps, he conducted an experiment on the spinal cord of a rabbit. His assistant stood by as Bell's hands moved with delicate precision, severing the anterior portion of the cord. The rabbit's limbs fell still. Bell leaned in, breath catching as he studied the response. He repeated the process, this time severing the posterior section. The reaction was immediate and different.

The results confirmed what had long been suspected but never proven: the anterior portion of the spinal cord governed motor function, while the posterior controlled sensation.

The implications were significant. Bell had uncovered a fundamental truth—one that would form the cornerstone of clinical neurology.

By 1815, Bell's work had earned him accolades, but the call of duty was never far away. News of the Battle of Waterloo reached London like a thunderclap, shaking the city with tales of unimaginable carnage. Bell began preparing for what lay ahead. He packed his surgical tools and, as always, his sketchbooks. The memories of Corunna lingered in his mind, the cries of the wounded, the chaos of the docks, but he knew this would be worse. Far worse.

As he boarded the ship to Brussels, Marion pressed his hand, her face a mask of quiet concern. "Take care of yourself," she said softly. Bell nodded, though he wasn't sure how to truly prepare for the horrors of war.

Twelve days after the bloody Battle of Waterloo, Charles Bell stepped through the doors of the *Gens d'Armerie Hospital* in Brussels. The stench struck him first—a suffocating blend of blood, rot, and sweat that clung to the air and seeped into his lungs. The groans of the wounded were ceaseless, rising and falling in a grim symphony that echoed along the overcrowded corridors. Every corner of the makeshift hospital seemed to sag under the weight of suffering. Soldiers lay crammed together on cots, stretchers, even the cold stone floor—some staring into nothing, others twisting in agony. Rats scurried undisturbed, their claws clicking across blood-slicked boards.

The scene was overwhelming, but Bell forced himself forward. There was no room for hesitation—not here, not after what had unfolded at Waterloo.

The battle, fought on the 18th of June 1815, had lasted a single day, but left behind decades of scars. It was the final, thunderous blow in Napoleon Bonaparte's long and bloody campaign. After his escape from exile, he had sought to reclaim Europe through one last show of might. But the Allied forces—British under the Duke of Wellington, and Prussian under General Blücher—had brought him to ruin. The price of peace had been high. Over 55,000 men lay dead or wounded in the fields of Belgium, their bodies broken by cannon fire and musket shot. Victory had been won, but the cost was almost beyond reckoning.

For Bell, the hospital in Brussels was the war's grim epilogue. Most of the British wounded had already been treated and sent home. Those who remained were French, left behind, abandoned in the shadow of defeat. He had known it would be terrible. Even so, nothing could have prepared him for this.

He passed a young Frenchman on a stretcher, his chest rising and falling in shallow, ragged breaths. The man's uniform was little more than tatters, and beneath it, a jagged wound split his side—raw, inflamed, and filthy. Bell knelt beside him, pulling back the sodden cloth with care. The soldier flinched but made no sound, either too weak or too far gone for protest.

"You'll be all right," Bell murmured, though he knew the words were hollow. His hands moved steadily, cleaning the wound with what little alcohol remained and applying fresh bandages. Around him, the ward buzzed with groans, shouted orders, the clatter of surgical tools but he shut it all out. Focus was the only way to endure.

For three days and nights, Bell worked without rest. His scalpel glinted in lantern light as he amputated limbs too ruined to save, stitched wounds that reopened with each cough or convulsion, and fought—often in vain—to slow the spread of infection. The hospital teemed with disease. Flies swarmed over exposed flesh, and the air was thick with the cloying reek of decay. For every life he saved, another slipped beyond reach, claimed by fever or sepsis. The losses gnawed at him; each one a silent indictment of what medicine could not yet do.

During a rare lull, Bell leaned back against the cool stone wall, his body sagging. He closed his eyes, but rest did not come. Behind his lids: the battlefield. Shattered limbs. Torn flesh. Faces twisted in agony. The ghosts of the wounded followed him. This was the cost of glory. This was the price of history.

He opened his journal and began to write, his hand trembling faintly. One case refused to leave him—a soldier whose humerus had been shattered by a musket ball, the shot continuing its path through the ribs. The decision had been made to amputate at the shoulder, but the man had bled out before the procedure could be completed. Bell scrawled:

> *"The ball struck the head of the humerus and shattered it, passed through and wounded a rib. It was resolved to amputate at the shoulder joint. It was reported to me that the patient sunk from loss of blood. I thought myself entitled to say that the method followed by our army surgeons was too bold, and not suited to common practice, and especially in a case like this, when the patient was reduced by a complication in the wound."*

The criticism was not directed at one man—it was aimed at the system. A system that prized speed over subtlety, boldness over caution. A system that left men to die in filth.

Later that day, as Bell was halfway through another amputation, the pressure cracked.

"You're too cautious, Charles. These men don't have time for second-guessing," snapped Robert Knox, his assistant, voice sharp enough to draw glances.

Bell straightened, jaw clenched. "Caution is the only thing keeping them alive, Robert. Rushing through an amputation without considering the patient's condition is reckless."

Knox's expression hardened. "Reckless? Look around you, Charles. How many of these men have died because we waited too long?"

Bell set down his scalpel. "And how many would have died regardless? These men arrived half-dead. The infection, the filth—it's not just what we do with the knife, and you know it."

Knox didn't respond. The tension lingered. Bell pushed it aside and returned to his patient, his hands moving with the grim determination of a man who refused to stop, even as his body screamed for rest.

When the ward settled once more, Bell reached for his sketchbook. It grounded him. Amid blood and screams and loss, the act of drawing offered a flicker of clarity. Sitting on an upturned crate, his hands still streaked with blood, he began to sketch. The soldier with the chest wound— the laboured rise and fall of his breathing, the tension in his clenched jaw, the desperate, defiant light in his eyes.

Watercolour of Wounded Soldier with Open Chest Wound, Head and Right Arm Bandaged, Waterloo, 1815 by Charles Bell

A guttural cry split the air, jagged and desperate. Bell turned, his eyes scanning the crowded ward until they landed on a British soldier lying on a blood-soaked cot. The man clutched at the space where his left arm had been, his face pale, his skin sheened with sweat and streaked with grime. Dirt had settled into the creases around his mouth and eyes, giving him the ghostly look of someone who had crawled out of the grave. Without hesitation, Bell hurried over and knelt beside him.

The soldier's arm had been torn off by an exploding shell—an injury so violent and final that most never survived it. A crude tourniquet had been wrapped tight around the upper limb, likely applied by a field surgeon or a fellow soldier amidst the confusion. The fabric was stained deep brown with dried blood, the skin beneath angry and swollen, the wound's edges already beginning to rot in the hospital's foul heat.

Bell studied it carefully. The ligature had done its job. It had stopped the bleeding long enough to bring the man here, to this stinking ward. That in itself was remarkable. Most men with such injuries died where they fell, bled out in the mud before help could reach them. The fact that this man had survived was a minor miracle.

"Hold on, soldier. You're going to be all right," Bell said softly, his tone steady despite the chaos around him.

The soldier turned his head, his dark eyes wide with pain and fear. "Doctor," he gasped, his voice barely audible. "The pain... it's unbearable."

Bell gently touched the man's uninjured shoulder and offered a faint smile. "I know," he said gently. "You are strong. Stay with me a little longer. We'll get you through this."

He examined the binding more closely. The tourniquet, though lifesaving, had cut too deep. The tissue around the wound was dying. Gangrene wasn't far behind. In a ward like this, with no fresh air and precious few clean bandages, the man's chances were shrinking by the hour. He needed more than what this place could offer. He needed London. He needed time.

Bell looked the soldier in the eye. "We're going to get you out of here," he said quietly. "To London. This wound needs more than I can give you in Brussels."

The man's lips quivered, his hands trembling slightly as he gripped the rope beside him—used to manoeuvre himself when he still had the strength. "Thank you, sir," he whispered. "Please... don't let me die."

Bell stood, calling out for an orderly. "Prepare this man for transport. He's stable enough to make the journey."

The orderlies moved quickly, lifting the soldier with rough tenderness, their own uniforms soaked through with blood and sweat. As they worked, Bell sat nearby on the edge of a cot, the familiar weight of his sketchbook in his lap. He opened it, the spine crackling slightly, and drew.

His hand moved across the page, swift but precise. The twisted ligature, the torn skin, the tight lines of pain around the soldier's eyes—he captured it all. This wasn't just a habit. It was a necessity. In the act of drawing, Bell could find order amidst the chaos, sense amidst the senselessness. The sketch was not just a record of anatomy; it was a testament to survival. A way to ensure this moment—this soldier—would not be forgotten.

As he worked, he glanced up, catching the soldier watching him with a mix of curiosity and dread. Bell offered a small nod. The soldier didn't smile, but something in his expression softened—a flicker of trust, or perhaps resignation.

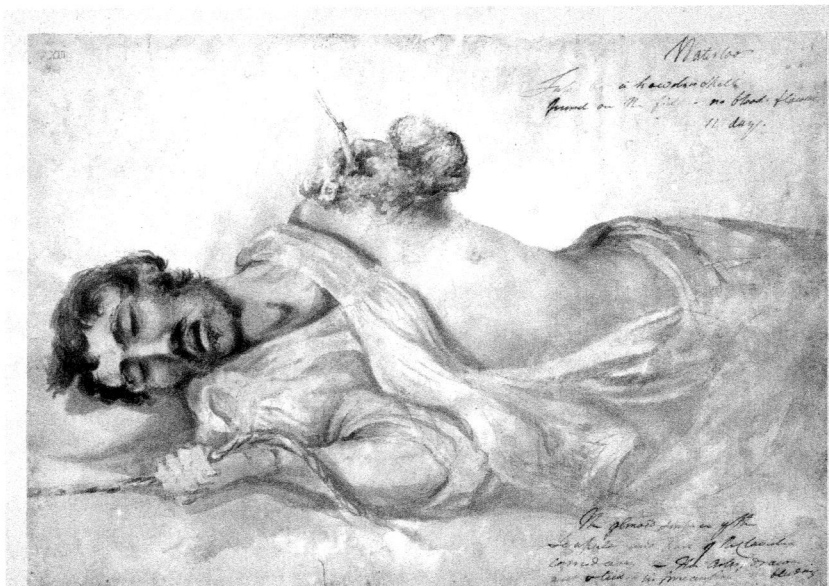

Watercolour of Wounded Soldier with Left Arm Torn Off by Exploding Shell, Waterloo, 1815 by Charles Bell

When the orderly returned, ready to transport the soldier to a wagon bound for London, Bell stood and closed his sketchbook. He looked down at the man who had fought so hard to hold on.

"What's your name, sir?" the soldier asked weakly as he was lifted onto the stretcher.

Bell hesitated for a moment. "Charles Bell," he said simply.

The soldier gave a faint nod, his lips curling into the ghost of a smile. "Thank you, Charles Bell."

Bell watched as the stretcher was carried away, his expression stoic but his heart heavy. He had done what he could, but he knew the man's survival was far from guaranteed. Still, sending him home felt like a small victory—a single light in the overwhelming darkness.

His gaze drifted back across the ward. His work was not yet done.

In the far corner, a soldier sat propped against the stone wall, silent and still. Bell walked slowly towards him, noting the slackness in the man's face, the blank stare fixed somewhere in the middle distance. His breathing was shallow, his skin flushed, a faint tremor pulsing through his jaw.

The man had taken a musket ball to the head.

Bell knelt, examining the damage. A crude, cross-shaped incision had been made in the scalp—evidence of an earlier attempt to relieve pressure or remove bone fragments. It had been necessary. It had also opened the door to infection. Now, pus and fluid wept from the wound, and a portion of brain tissue had begun to extrude.

The soldier didn't respond to Bell's touch. His gaze didn't flicker. His body was still fighting to stay alive, but the man himself had likely already left.

Bell knew the signs. Meningeal irritation. Infection setting into the brain's delicate lining. He swallowed hard, heart tight. He could do little. No surgeon could. But still, he reached for his sketchbook.

If he could not save the man, he could at least bear witness.

He drew slowly, the pencil gliding across the page. The cross-hatched wound. The swelling. The dull, distant expression. Each line captured a truth too often lost in statistics and reports. This was the cost of war, rendered not in numbers but in flesh and sorrow.

As he sketched, he paused and looked up, meeting the soldier's vacant eyes. For a moment, Bell felt as though he were staring into the heart of suffering itself—an unspoken howl of anguish behind that hollow gaze. He lowered his pencil. Then, slowly, he continued.

Six days later, the soldier died—claimed by the infection Bell had known would come.

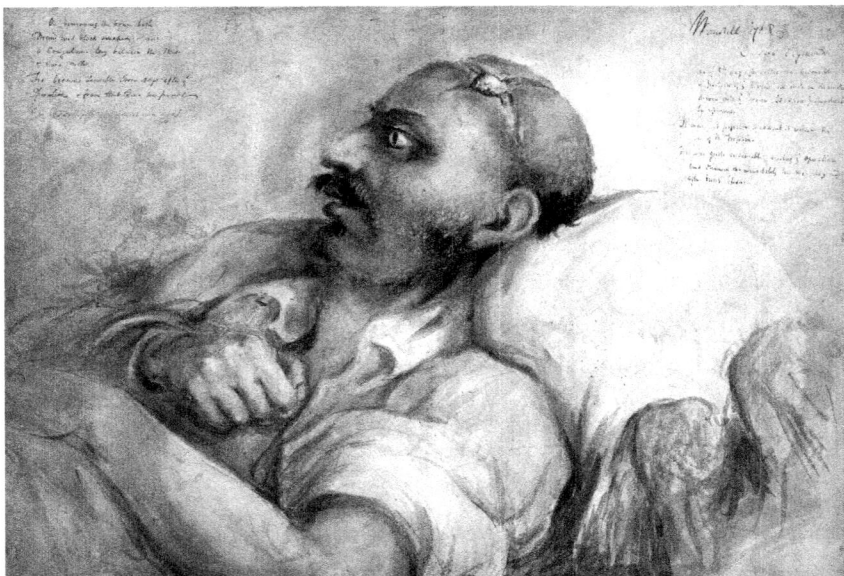

Watercolour of Wounded Soldier with Serious Gunshot Wound to the Skull, Waterloo, 1815 by Charles Bell

By the eighth day, Bell was exhausted. Not just from the endless hours, the amputations, or the filth, but from the weight of helplessness. Of the 12 major surgeries he had performed, only one patient had survived. The rest were gone, consumed by fever, shock, or the quiet inevitability of infection. The losses hung over him like a shroud, each face etched behind his eyelids, each name unspoken but not forgotten.

There were moments—brief, bitter moments—when he questioned whether anything he had done had made a difference. Whether his hands had healed or merely delayed the dying. But he pressed on. Not because he was certain, but because he could not bear the thought of stopping.

He boarded the ship back to London in silence. The wind was sharp, the sea grey and restless, and Bell stood at the rail as the coast of Belgium receded behind him. His coat was still stained from the hospital. He hadn't had time to change.

When he arrived home, Marion met him at the door. She said nothing, simply took his coat and led him inside. The house felt too quiet after the hospital, the silence almost hollow. It was only later, once he'd washed and changed, that he found the letter.

It had been left on his writing desk, addressed in the careful hand of a colleague.

He opened it with fingers still sore from work, unfolding the page with a quiet apprehension.

"The patient has made an excellent recovery," the letter read. *"He expressed his deepest gratitude to you for saving his life."*

Bell read the words again. Then once more.

It was the soldier—the young man with the torn-off arm, the one he'd sketched before sending home. Against the odds, he had lived.

Bell leaned back in his chair, the paper still in hand. His lips pressed into the faintest of smiles. It wasn't triumph. It wasn't even relief. It was something quieter. A fragile kind of hope.

One life. One survivor.

In a sea of death, it wasn't much. But it was something. And sometimes, in the long shadow of war, something was enough.

Even years after the chaos of Brussels, Charles Bell's fascination with the human body remained undiminished. If anything, it deepened. To him, every nerve and muscle told a story—every tremor, every spasm, every flicker of the skin, a clue to the hidden language of the body. What had once been a surgeon's discipline became something more expansive: a lifelong pursuit to understand how flesh and feeling intertwined.

His interest gravitated toward the nervous system, that delicate web of threads linking sensation to thought, action to meaning. It was a realm that, in his time, was poorly understood—a mystery bordered by superstition and speculation. But to Bell, it held the key to everything. Not just how we move, but how we feel. How we express. How we *are*.

He began with the eye.

It had fascinated him since youth—its intricacy, its elegance, its strange and silent eloquence. In his 1823 papers, Bell described its astonishing anatomy with reverence. Six of the nine known cranial nerves, he observed, were directly involved in its function. They worked in concert, an orchestra of motion and sensation, allowing the eye to track, to blink, to focus, to respond.

But for Bell, the study of the eye was not an end in itself. It was the gateway to something more human. Something expressive. The same nerves that allowed the eye to move also governed the way we reveal fear, hope, suspicion, amusement. The muscles of the face were not just mechanical—they were emotional instruments.

This was never clearer to him than during the war, when the silent expressions of wounded soldiers often spoke more than words ever could.

It was this insight that drew him deeper into the mysteries of facial nerve function.

In his clinic, Bell began to see patients suffering from an agonising condition: facial neuralgia. A searing, unpredictable pain that left them stricken, twisted, unable to express themselves. Some had suffered trauma. Others, infection. But all shared the same loss— their faces no longer obeyed them.

Surgeons of the time, desperate to relieve the pain, often severed the facial nerve entirely. Sometimes it helped. Sometimes the agony lessened. But the cost was high. Patients were left with immobile features, half their faces slack and frozen. They couldn't smile. Couldn't frown. Couldn't *be* themselves. It was more than disfigurement. It was erasure.

Bell was unsettled. To him, the face was not merely a collection of muscles. It was the body's first voice. The most honest canvas of the soul.

One day, a man came to his clinic with a grimace frozen on his face. The left side hung slack, the cheek and brow immobile. The right side, by contrast, was contorted, overcompensating, pulled into an almost grotesque expression. His left eyelid wouldn't close, leaving the eye dry and wide. Unprotected.

"Dr Bell," the patient began, his voice edged with fatigue and frustration, "my left eyelid won't blink, and my cheek feels like it's made of stone. It's as though half my face has abandoned me."

Bell studied him intently. He asked the man to smile. To raise his eyebrows. To shut his eyes. With each request, the symptoms revealed themselves: the left side refused to respond. When the man tried to wink, the eye rolled upward, revealing the white of the globe beneath the lid.

Bell nodded and noted his findings in careful, precise script:

> *"The face is twisted to the right side. The left nostril does not move in respiration. The eyelids of the left side are not closed when he winks, although when he attempts it, the eyeball is turned up. The cheek is relaxed, and the forehead on the left side unruffled."*

That upward roll of the eye caught his attention.

It was subtle. Easy to miss. But Bell recognised it as a clue— a reflex that became known as *Bell's phenomenon*. He explained it as a primitive protective mechanism, where the eyes instinctively roll upward when the eyelids are suddenly shut. Although this reflex occurs in everyone, it becomes more noticeable in cases of facial paralysis when the eyelid cannot fully close.

Bell continued to study these cases, and over time, his insights became clearer. He began to distinguish between two types of facial paralysis. Peripheral paralysis, caused by damage to the facial nerve after it had left the brain, usually affected the entire side of the face—from forehead to chin. Central paralysis, by contrast, originated from within the brain itself and typically spared the upper face. The differences were subtle—but vital.

His work on facial paralysis would later be immortalised in the condition known as *Bell's palsy.* However, for Bell, the naming of the disorder was secondary to its significance. What mattered to him was that understanding the facial nerve had the potential to improve lives.

His influence spread far beyond the operating theatre. His studies on facial expression, pain, and emotion later inspired Charles Darwin, whose landmark work *The Expression of the Emotions in Man and Animals* drew directly from Bell's writings. Darwin recognised what Bell had long known: that the face was the mirror of the soul and that every wrinkle, twitch, and smile carried an element of human experience.

Despite the acclaim that followed; his knighthood by King William IV in 1831, his celebrated lectures, and his role in founding the Medical School at the *University of London*, Bell never stopped working. He returned to Edinburgh later in life, where he continued to teach, sketch, and probe the human body's mysteries. He believed that medicine and art were not separate disciplines. His anatomical drawings were more than scientific tools; they were portraits of humanity. For him, every nerve, every muscle, every line drawn on paper was a bridge between art and medicine.

What is Bell's Palsy?

Bell's Palsy is a condition that causes sudden, temporary weakness or paralysis of the muscles on one side of the face. This results in the affected side appearing drooped or unable to move normally. The condition occurs when the facial nerve controls these muscles and becomes inflamed or compressed—often due to a viral infection or other irritation.

While its onset can feel alarming, *Bell's Palsy* is usually not linked to any serious underlying health issues. Most people recover fully within weeks to months, often with the help of treatment like steroids or physiotherapy. For many, patience and care are key, as the body's natural healing process restores strength and movement to the affected muscles.

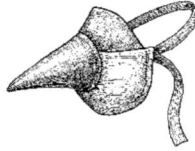

CHAPTER 13

The Dancing Curse

EAST HAMPTON ON Long Island during the late 1850s was a village of quiet contrasts. Beneath its rustic charm and serene landscapes—sprawling woods, weathered fences, the steady lull of Atlantic waves—lay an undercurrent of unease. Whispers of a strange affliction moved through the community like ghost threads, stitching together tales of misfortune and dread. To those who had seen it, it was simply known as *"the disorder."*

The Huntington family, respected for generations of medical service, had long stood at the centre of village life—both witnesses to its troubles and stewards of its care. George Huntington, born in 1850, was the third in a distinguished line of physicians. His grandfather, Abel Huntington, had laid the foundation of a thriving practice, which his father, George Lee Huntington, continued with tireless devotion. From early childhood, George rode beside his father on visits, a quiet observer with wide, curious eyes, absorbing the craft of healing one patient at a time. The roads of East Hampton often echoed with the creak of their carriage wheels as they travelled from home to home, offering what comfort and treatment they could.

And yet, even amid the rhythms of everyday life, the peace of the village was sometimes broken by tragedies that left a deeper kind of silence. One such story belonged to the village blacksmith. Once admired for his strength and steady hands, he had become a figure of pity—plagued by uncontrollable jerking movements, sudden grimaces, and a slow, merciless unravelling of self. One morning, the villagers found him hanging in his workshop. He could no longer bear the humiliation of *the disorder* and had chosen death over the long road of suffering that lay ahead. He wasn't the first. For others like him, the icy pull of the sea offered a cruel, final escape. Their stories—of bodies pulled ashore or never found—drifted through the village like half-remembered prayers.

It was against this sombre backdrop that George Huntington encountered the moment that would change everything.

It was autumn, and the trees had begun to turn—leaves glowing with fire and gold as he rode beside his father on one of their regular visits to Amagansett. The road wound through dense woodland, the air cool and quiet, the soft rustle of leaves soothing and familiar. But that calm was shattered as they rounded a bend.

By the roadside stood two women—a mother and daughter—moving as if gripped by some invisible force. Their bodies jerked and twisted in grotesque rhythm, faces frozen in grimaces that seemed carved from pain. George stared, rooted to his seat, unable to look away. The movements were unnatural, like a cruel parody of dance, as though the women had been turned into puppets by some torment within their own flesh.

His pulse quickened. A knot of fear and fascination twisted in his chest. *Why are they moving like that?* It was horrifying and somehow mesmerising. As if he were witnessing something sacred and profane all at once.

He turned to his father, expecting alarm, but George Lee Huntington remained composed. His brow furrowed with quiet concern, but his voice was steady as ever. Without a word, he stepped down from the carriage and approached the women—not with panic or revulsion, but with compassion. He spoke to them gently, treating them not as curiosities or victims, but as human beings who deserved care.

George watched in silence, a thousand questions crashing through his mind. He had seen illness before but nothing like this. These women seemed possessed by something cruel and unstoppable. Their expressions haunted him.

When his father returned to the carriage, George could no longer hold back.

"What's wrong with them?" he asked, his voice cracking into the quiet.

His father sighed. Years of sorrow moved behind his eyes. He spoke slowly, explaining what he could—the strange and persistent nature of *the disorder*, its grip on body and mind alike. George listened, but the words only deepened the imprint of what he'd seen. The image of those women—twisting, grimacing, caught in their own bodies—would never leave him.

The disorder was also known as *chorea*. The word itself came from the Greek *choreia*, meaning "dance"—a term that conjured the strange, restless movements that plagued its sufferers. In centuries past, it had gone by another name: *dancing mania*. That, too, was no exaggeration. Long before science offered any explanation, chorea swept through European villages like wildfire, blurring the line between disease and possession, between fear and folklore.

During the Middle Ages, when Europe was convulsing beneath the weight of the bubonic plague, these outbreaks became part of the collective nightmare. At times of mass fear and grief, when reason faltered and desperation took root, crowds would gather—men, women, even children—gripped by an unstoppable force. Their bodies jerked and spasmed as though seized by invisible strings. They danced, not in joy, but in torment, limbs flailing, feet pounding, eyes wide with terror or emptiness. They could not stop. Some collapsed from exhaustion. Some died where they stood.

The first recorded incident took place in the now-lost Saxon village of Kölbigk, where, in the shadow of a church, 18 villagers began to dance uncontrollably. The priest, horrified by what he saw, declared them cursed. He ordered the church doors shut against them, sealing them out. No prayers. No absolution.

Over the centuries, these outbreaks would appear again and again, rippling through communities already on edge. Superstition filled the gaps where medicine had no answers. People spoke of divine punishment, of purgatory leaking into the world of the living. The afflicted were said to be mocking the saints, dancing in defiance of heaven itself.

Desperate for relief, sufferers—and their families—sought salvation where they could. Pilgrimages were made to sacred sites, where the sick prayed for a miracle. They believed healing could be found in the bones of saints, in the echo of hymns, in the touch of sanctified stone.

One name rose above all others: Saint Vitus.

Vitus was a boy when he died—a Christian martyr from Sicily, persecuted during the brutal reign of Emperor Diocletian in the early fourth century. According to legend, Vitus was the son of a Roman senator, but he rejected the pagan faith of his family and secretly embraced Christianity. When his defiance was discovered, he was arrested and subjected to a series of tortures. One of the most enduring tales describes him being plunged into boiling oil— yet miraculously surviving, his body untouched by the heat, his faith unbroken. He was eventually executed, but his legend only grew.

In the centuries that followed, stories of his miraculous endurance spread across Europe. One in particular told of Vitus curing the son of Emperor Diocletian, who had been suffering from uncontrollable convulsions—possibly epilepsy. Brought before the emperor,

Vitus is said to have laid hands upon the boy and prayed over him, whereupon the fits ceased. Because of this, Vitus came to be seen not only as a symbol of faith under fire, but as a heavenly protector of those suffering from seizures, tremors, and inexplicable bodily afflictions. By the 15th century, he had become the patron saint of dancers, epileptics, and all those tormented by involuntary movement. His name became permanently bound to the condition that seemed to haunt Europe in waves: *Saint Vitus' Dance.*

In the Alsatian town of Zabern, pilgrims gathered at the chapel built in his honour. They came on foot, some limping, some carried, some still shaking. They touched the cold stones of the church walls, whispered prayers through chattering teeth, and left offerings at the altar. Some claimed to have been healed. Others did not return at all.

Medicine began to change in the centuries that followed. The Enlightenment dimmed some of the superstition, though not all. By the 19th century, science had turned its eye to chorea, trying to separate myth from mechanism.

In Paris, Jean-Martin Charcot took up the study. Charcot gave chorea the weight of academic legitimacy, though even he struggled to untangle its many forms. He saw it as a constellation of symptoms—sometimes hereditary, sometimes triggered by illness or trauma. But he could not yet draw a clear line between its causes. The difference between autoimmune and genetic chorea remained hidden, just beyond reach.

That would be left for another to uncover.

Still, the hereditary thread had long been woven into the tapestry of the disease, even if no one had yet found the pattern. Long before microscopes and gene theories, it passed silently through families—crossing oceans, planting itself in new soil. In 1630, two brothers from Suffolk, England, emigrated to the Massachusetts Bay Colony. Among the luggage and livestock, they carried something unseen:

the genetic mutation that would one day be known to medicine, but for generations remained a silent curse.

Their descendants flourished. Families grew. But with growth came suffering. Children inherited more than names—they inherited uncontrollable twitching, grimacing, and jerking movements. In tight knit, deeply religious communities, these symptoms were misunderstood. Their twisted expressions and sudden movements were read not as illness, but defiance. An affront to God.

In some villages, people whispered that they were mocking the crucified Christ—grimacing like demons, writhing like the damned.

And in Salem, those whispers turned deadly.

During the infamous witch trials, symptoms of chorea—uncontrollable movements, contorted faces, disjointed speech—were held up as proof of possession. Girls who twitched and shrieked were dragged before tribunals. Their affliction, misread through the lens of fear and piety, became a death sentence. Some historians now believe that the hereditary disorder passed through Puritan bloodlines played an unspoken role in the panic that consumed Salem.

In East Hampton, far from those infamous trials but still tethered to their legacy, the disorder continued its quiet devastation. One of its most haunted families was the Hedges clan—one of the oldest and most respected names in the region.

On the morning of 12th June 1806, Captain David Hedges awoke to find his wife missing. Panic set in quickly. The villagers searched the fields, the woods, the shoreline. The air was heavy with the dread of what they might find. It was her mind, they knew. It had begun to go.

Hours later, the sea gave up her body.

She had walked into the Atlantic rather than face what was to come. Her mother had done the same.

The Hedges family had long been under the care of the Huntington doctors. Abel Huntington, George's grandfather, had watched the disease pass from generation to generation, tracking it not as a scholar, but as a witness. The bond between doctor and family was one of quiet trust. By the time George was old enough to accompany his father on house visits, the stories of families like the Hedges had already left their mark.

He saw it in their eyes. In the way they moved. In the way others avoided their gaze.

Moments like these gave shape to his determination. The day he saw the women by the roadside—twisting, grimacing—it did not begin something in him. It confirmed what had already taken root.

He would not look away.

George Huntington's journey into the medical world began at the *College of Physicians and Surgeons of Columbia University*. He graduated in 1871, after years of demanding, often exhilarating study—filled with late nights, complex dissections, and moments of revelation that made the long hours worthwhile. For Huntington, medicine was never just a profession. It was a lineage. The weight of his family's legacy—the steady hands of his father, the quiet wisdom of his grandfather—was always with him. He had inherited not only their vocation, but their long, careful observations of a particular affliction that haunted their community.

After completing his studies, Huntington returned to East Hampton, bringing with him the formal knowledge of medicine and the restless curiosity that had first stirred in him as a boy. He buried himself in the old records his father and grandfather had kept— leather-bound volumes filled with detailed notes, inked diagrams, and generations of insight. These documents were more than case studies. They were the living history of the disorder. Decades of patient encounters, sketches of symptoms, even personal reflections, lined the pages.

At night, Huntington would pore over these volumes in the flickering light of his lamp, the room hushed but for the turning of pages. He read slowly, absorbing every detail, every family name, every note in the margins. The faces of the afflicted—so many of them familiar—seemed to rise from the pages. It was as though he could hear their movements again: the shuffle of uncertain steps, the sharp gasp of an involuntary spasm, the murmurs of families waiting in the next room.

His father remained close during this time—not just as a mentor, but as a collaborator. Together, they reviewed the notes, cross-checked observations, and talked late into the evening about what they'd seen. The elder Huntington corrected his son gently, pointing out subtle patterns or clarifying points of family history. These conversations became a kind of apprenticeship all their own—half science, half storytelling.

What emerged from these long nights was not just a set of observations, but a clarity that had evaded even experienced physicians. George Huntington wasn't just inheriting the work—he was preparing to make sense of it.

That moment came on 15th February 1872, when, at just 22 years old, he stood before the *Meigs and Mason Academy of Medicine* in Middleport, Ohio, to present his findings. The night before, he barely slept, his mind racing with anticipation and doubt. Would his findings be met with interest, or would they be dismissed? He replayed his arguments in his head, refining the words he would use to bring clarity to a disease that had brought so much suffering.

The hall was filled with physicians. Some sat with arms folded, sceptical. Others leaned forward, curious. As Huntington rose to speak, his voice was steady, but there was tension behind it—an urgency that came not from ego, but from lived experience.

He did not speak from theory alone. He drew from what he had seen with his own eyes. He spoke of the families he had known, the symptoms passed down like heirlooms, the devastating, inescapable grip of the disease. He described how it began—subtle twitches, awkward gestures—before escalating into violent, uncontrollable movements that left the body bent and broken. He spoke, too, of the mind—how it, too, unravelled—giving way to dementia, paranoia, even madness.

Huntington defined three key characteristics of the disease. First, its hereditary nature. If a parent had symptoms, at least one child would inevitably inherit the condition—provided they lived long enough.

> *"Unstable and whimsical as the disease may be in other respects," he told the audience, "in this it is firm: it never skips a generation to again manifest itself in another; once having yielded its claim, it never regains it."*

In other words: if the disease failed to appear in a child, it disappeared from that bloodline forever.

Second, he emphasised the psychological toll. Those afflicted were often overwhelmed by despair. Suicide was not uncommon. Families lived in fear—not only of the disease itself, but of the stigma it carried.

Finally, he spoke of its unforgiving nature. Once it began, it progressed without pause—claiming both body and mind.

> *"As the disease progresses," he explained, "the mind becomes more or less impaired, in many amounting to insanity, while in others, mind and body both gradually fail until death relieves them of their sufferings. When once it begins, it clings to the bitter end."*

His words carried not just clinical insight but a deep empathy for the patients and families devastated by the disorder. The room was silent as the audience absorbed the gravity of his observations, their faces reflecting both fascination and sorrow.

Huntington shared case studies to bring his findings to life, recounting stories of individuals who had lost their sanity and eventually their lives, of communities ostracising those whose bodies betrayed them with erratic movements and grimaces.

Yet, Huntington's tone remained humble. As he concluded his lecture, he remarked, *"I know nothing of its pathology and offer my account not that I consider it of any great practical importance to you but merely as a medical curiosity, and as such, it may have some interest."*

The room erupted in applause as Huntington stepped away from the podium, the pressure of months of preparation finally lifting. Relief and satisfaction washed over him. He had done it; he had shared his family's legacy and his discoveries with the medical world, and he knew he had made an impact.

Shortly after, he submitted the manuscript to the *Medical and Surgical Reporter of Philadelphia*. It was published on 13th April 1872, and in just a few pages, Huntington had achieved something rare: clarity. The medical community took notice. His observations— both precise and compassionate—marked the beginning of a new understanding. What had once been dismissed as madness or moral failing was now recognised as a disease: *Huntington's chorea*. In time, it would simply be known as *Huntington's disease.*

But beyond the diagnosis and the name, there was something else. Huntington had made the suffering visible. He had shown that behind the symptoms were human lives—worthy of study, yes, but also of dignity.

Despite the promising start to his medical career, George Huntington chose a different path. While many of his peers chased academic positions or devoted themselves to research, he followed the quieter road walked by his father and grandfather. He became a country doctor. Not for lack of ambition, but because he believed, with quiet conviction, that his lecture on chorea had fulfilled his responsibility to science. The rest of his life, he felt, belonged to his patients.

In 1874, he married Mary Elizabeth Hackard. Together they returned to East Hampton, though the tranquillity of the village would not hold them long. Before the year was out, the couple moved to LaGrangeville, New York—a small rural community surrounded by fields and dense forest. It was the kind of place where neighbours knew each other by name, where life moved at a gentler pace. Huntington saw in it not remoteness, but possibility. A chance to build a life rooted in purpose and presence.

Here, he established a thriving medical practice. His days were filled with house calls, the steady rhythm of horse hooves on dirt roads, and the weight of a satchel filled with well-worn tools. He travelled through woods and weather to reach patients in need—never rushed, never indifferent. In this work, he found satisfaction. Not the kind that filled journals or drew accolades, but the quiet kind that came from being useful, day after day.

He was more than a physician. He was a neighbour, a confidant, a man who knew the names of children and the worries of their parents. When families could not afford to pay, he treated them anyway. Payment, when it came, might be a sack of potatoes, a brace of ducks, a warm loaf of bread left on the doorstep. Huntington never minded. Medicine, to him, had always been a form of service, not a transaction.

Away from his practice, he lived a life rich in simple pleasures. He was an avid outdoorsman, often escaping into nature with his rifle and sketchbook. He had a keen eye for wildlife, and his drawings of game birds captured their delicate grace with the same attention he once gave to the study of nerves. Music, too, filled his home. He played the flute, while Mary accompanied him on the piano. Their melodies drifted through the house on quiet evenings, softening the edges of a long day's labour.

Those who knew him spoke of his humour—gentle, dry, and precise. He had a way of defusing tension with a word, a smile, a well-placed line. He did not posture. He did not preach. But he listened, and people trusted him for that. His wit was never cruel, and his kindness never performative. It was simply who he was.

Yet his life was not untouched by difficulty. Huntington suffered from chronic asthma, a condition that often left him breathless and fatigued. In 1901, seeking relief from New York's pollution, he moved south to Asheville, North Carolina, drawn by the promise of fresh mountain air. For two years, he lived at a slower pace. He wrote letters, sketched birds, and walked beneath the trees. Though the move interrupted his medical work, it did not dampen his spirit.

One summer afternoon, in the midst of another asthma flare, he crafted a makeshift respirator out of pasteboard and gauze—a contraption that, once tied across his face, resembled the beak of a bird. He wore it out into the sunlight and was soon spotted by his neighbour, Mrs Losee.

Her eyes widened. "Doctor! What are you wearing that thing on your nose for?" she asked, half-laughing.

His eyes twinkled behind the absurd mask. "To keep it from poking into other folks' business," he replied.

Their laughter echoed through the warm air.

In 1903, with his health improved, Huntington returned to New York. He settled in Hopewell Junction, not far from LaGrangeville, where he resumed his practice and was welcomed like an old friend. He served as a visiting physician at *Matteawan General Hospital* and as health officer in the nearby town of Fishkill. His patients—many of them farmers and labourers—relied on his steady care, and he gave it freely, as he always had.

He continued to practise until 1915, when his health at last began to fail. At the age of 64, he stepped back, his body too weary to keep up the pace he had long maintained. Even in retirement, he remained a quiet source of strength to those around him.

On 3rd March 1916, George Huntington died of pneumonia, aged 65. He passed away in the home of his son, surrounded by family, his life drawing to a close with the same humility and quiet purpose that had shaped it.

His discovery—made in a single lecture hall on a winter evening— would ripple out through generations. But perhaps more than that, his legacy would lie in the way he saw people: not as curiosities, not as specimens, but as lives worth understanding, and worth helping, however one could.

What is Huntington's disease?

Huntington's disease is a rare, inherited condition that causes the gradual breakdown of nerve cells in the brain, leading to significant changes in movement, thinking, and behaviour. It is caused by a genetic mutation in the *HTT* gene, which produces an abnormal protein that damages brain cells over time.

The disease usually develops in adulthood, often between the ages of 30 and 50. Early symptoms may include subtle, involuntary movements, clumsiness, and difficulty concentrating. As the disease progresses, these movements become more pronounced, developing into uncontrollable jerking motions known as chorea. Alongside physical symptoms, individuals may experience emotional disturbances such as depression, irritability, or mood swings, as well as cognitive decline, which can impair memory, decision-making, and the ability to organise thoughts.

Huntington's disease is passed down in families through an autosomal dominant inheritance pattern. If one parent carries the faulty gene, each child has a 50% chance of inheriting the disease. If a person does not inherit the mutated gene, they cannot pass it on to their children, and the disease will no longer appear in that family branch.

While there is currently no cure for *Huntington's disease*, advances in medical research have improved our understanding of its genetic and biological mechanisms. Today, treatments focus on managing symptoms and improving quality of life, with medications available to reduce involuntary movements and address mood or psychiatric issues.

CHAPTER 14

In the Family

IN THE 1930S, New York City's Upper West Side was a place of sharp contrasts. The glamour of jazz clubs and new, clean-cut buildings stood in uneasy balance with the hunger and grit of the Great Depression. For many families, it was a daily struggle—life lived on a knife's edge. The streets were crowded, noisy, and unpredictable. Children played stickball until the clatter of arguments or the crack of fists sent them scattering. Apartments were packed so tightly that neighbours could hear the rise and fall of every whispered fight and muffled sob. In these close quarters, childhood was a short-lived privilege. Growing up wasn't gradual—it was abrupt.

Henry Thompson Lynch was born to an Irish family in Lawrence, Massachusetts, and found himself thrust into this boiling landscape when his family moved to New York. His father, once a travelling salesman with a comfortable income, had lost his job when the economy collapsed. His mother stepped into the gap, working long hours as a secretary just to keep the lights on. Their new apartment on the Upper West Side was small, loud, and always tense. Every expense was calculated. Every indulgence, no matter how minor, came with guilt. For Henry, that pressure was ever-present. He felt it in the air, in the silence that settled over the dinner table, in the clenched jaws and careful words of the adults around him.

Outdoors, things weren't easier. The streets demanded a kind of toughness, and Henry learned quickly how to meet that demand. Fights weren't unusual—they were expected. And for a wiry boy with a sharp tongue and something to prove, they became a fact of life. These weren't about winning or losing so much as about not backing down. He wasn't built for intimidation, but he developed a fire that others learned to respect. Still, there was no pride in it—just necessity. He fought because it was the only way to keep standing.

School, by contrast, felt irrelevant. What they taught in the classroom seemed far removed from the urgent lessons unfolding outside. At fourteen, Henry decided he'd had enough. He left school, not in protest, but with quiet certainty that whatever he needed to know wasn't written in textbooks. His real education, he believed, was waiting elsewhere—beyond chalkboards and bells.

And then came the war. It pulled at him like a tide, fierce and inescapable. Europe was in flames, and the Pacific was no less violent. For a boy already hardened by the city's sharp edges, the idea of fighting for something bigger than himself held an undeniable appeal. He was only 16 but desperate to join. When legal age stood in his way, he borrowed a cousin's identity, gave a false age, and enlisted in the Navy. By the time anyone discovered the truth, he had already proved he belonged.

He was assigned to a merchant marine ship and thrown straight into the storm of the Pacific War. What had once seemed distant was now terrifyingly real. As a gunner, he faced enemy fire, the deafening thunder of the ship's cannons, and the knowledge that every hour might be his last. During the liberation of the Philippines, he lost part of his hearing—a lasting injury, though one he rarely spoke of. What stayed with him more were the memories: the sea soaked in danger, the faces of friends he didn't come home with. But he endured, driven by loyalty, purpose, and something deeper still—a refusal to give in.

When Henry Lynch finally returned home, the streets of New York felt both familiar and strange. They had once been his entire world, a proving ground of fists and fire. But after the war, they seemed smaller and quieter, somehow. He wasn't the boy who had left. He was taller, stronger, and carried the weight of memories that had aged him beyond his years. He had seen too much—friends lost in moments, death brushing close more times than he could count. The city hadn't changed, but he had. What once felt like survival now felt hollow.

Adjusting to civilian life didn't come easily. The fight that had kept him going during the war hadn't left him; it had simply lost its direction. That restless energy needed somewhere to go, and he found it in boxing. The ring gave him something the streets and the sea no longer could—a focused outlet for the aggression, the frustration, and the drive that had shaped him. They called him *"Hammerin' Hank,"* a name that stuck, partly because it sounded good, but mostly because it fit. In the ring, he was ruthless, explosive, impossible to ignore. But strength has its limits. While his punches landed hard, his legs couldn't always carry him through the long bouts. No amount of training could fix what his body had absorbed over the years.

Accepting that his boxing career wouldn't take him where he'd hoped was a bitter pill. He had poured so much of himself into it—every setback, every surge of anger or fear transmuted into movement. Letting go of that dream felt like losing a part of his identity. But as the realisation settled in, something unexpected happened. He didn't break. Instead, he turned. The discipline, the resilience, the sheer force of will he had relied on for so long began to take a different shape.

Henry decided to return to school. He earned his high school diploma—a quiet triumph that spoke volumes about his determination—and started to look beyond the immediate. The rough edges of his past hadn't disappeared, but now, they pointed forward. He began to see knowledge not just as escape, but as a weapon—another way to fight, another kind of power.

Psychology and genetics drew him in, not because they were fashionable or impressive, but because they felt urgent. His own life had been a case study in survival, in the raw extremes of human resilience and vulnerability. On the streets, in the war, and even in the ring, he had seen people unravel and endure in ways that defied logic. Why did some collapse while others endured? What invisible forces shaped a person's ability to keep going when everything else had failed? These weren't abstract questions. They were personal.

Driven by this hunger to understand, he enrolled at the *University of Oklahoma* and graduated in 1951. But lectures and textbooks were just the beginning. He didn't want surface answers—he wanted to dig deeper. When he moved on to the *University of Denver* for a master's in clinical psychology, the focus sharpened. Psychiatric disorders like schizophrenia drew his attention. The mystery of them—the way a mind could come undone, seemingly at random—was both fascinating and haunting. He kept circling back to the same question: was it really random? Or was there something written in the blood, something inherited but unseen?

Lynch threw himself into the work. The same fire that had driven him on the streets and in battle now fuelled his studies. In this new arena, research was his fight, and knowledge his weapon. He wasn't content with theories that stopped at surface symptoms— he wanted to know what lay beneath. Each case study, each lecture, each late-night reading session became part of a larger quest to understand the hidden mechanics of the mind.

By the time he began his PhD studies in human genetics at the *University of Texas*, Lynch had found his trajectory. He believed, deeply, that the secrets to understanding not just psychiatric illness but the broader human condition were embedded in our DNA. He saw patterns where others saw coincidence, connections in places the textbooks overlooked.

But even as he immersed himself in the intricacies of chromosomes and hereditary markers, he sensed the limitations of theory alone. Science was essential—but sterile without purpose. He didn't want to live in the confines of academia, charting unseen variables with no direct link to the people suffering outside those walls. His aim wasn't just comprehension—it was change. He wanted his work to matter, to reach people, to help.

That realisation marked another pivot. If genetics could reveal the cause, medicine could intervene. Lynch realised that to truly help others, he needed to combine his growing scientific expertise with hands-on care. This led him to shift his focus toward medicine, marking the beginning of a new chapter.

By the time Henry Lynch earned his medical degree from *the University of Texas Medical Branch* in Galveston in 1960, he was no longer just a survivor or a student—he was a man with a mission. Still, even with his white coat and degree, something nagged at him. He hadn't come this far just to treat symptoms. He wanted to understand root causes, to close the gap between genetic theory and clinical practice. The term "medical genetics" was barely beginning to enter the lexicon, but Lynch was already living its principles.

His first steps into the medical world began with an internship at *St. Mary's Hospital* in Evansville, Indiana. There, the idealised image of medicine gave way to its messy, human reality. The corridors buzzed with urgency. Patients arrived with pain and fear etched across their faces. The workload was endless. But Lynch thrived. In the chaos, he found clarity. Each diagnosis, each chart, each bedside conversation refined his focus. The hospital wasn't just a place of treatment—it was a living archive of human suffering and resilience.

Next came a residency in internal medicine at the *University of Nebraska College of Medicine* in Omaha. Compared to the crowded cities of his youth, Nebraska felt vast and slow. But it offered something extraordinary: stability. Families here often stayed in one place for generations, and with that continuity came a rare gift—genealogical clarity. Medical histories weren't lost in transit or buried in bureaucracy. They were preserved, handed down, remembered. To Lynch, these communities weren't quiet—they were rich with data, living laboratories where inherited disease could be studied in full.

It was here, in 1961, that he established his first medical genetics clinic. Modest in scale but visionary in purpose, the clinic operated with little fanfare and fewer resources. Yet its work was groundbreaking. Lynch's enthusiasm proved contagious. Medical students, dentists, and colleagues across disciplines were drawn to the effort.

Their days were long and their cases often harrowing. But when they began plotting the regional distribution of hereditary diseases across Nebraska, something shifted. The maps they created weren't just scientific—they were personal, charting a generational inheritance of suffering that had, until now, gone largely unrecognised.

Still, there were nights when Lynch sat alone with the records, uncertainty pressing in. Were the patterns real, or were they the product of selective perception? He couldn't be sure. But something inside him insisted: there was more to uncover.

It wasn't until Lynch met a seemingly ordinary patient that the true scope of his work began to unfold.

It was a slow afternoon in 1962 when Dr Henry Lynch was asked to examine a new patient on the ward. Nothing unusual at first glance—just another a middle-aged man with the unmistakable signs of alcoholism, his body sagging into the hospital mattress as though it had finally given up trying to hold itself upright. The man's skin had the sallow tone of someone who'd been drinking too long, too often. His breath was reeking of stale liquor, his eyes rimmed in red. Lynch had seen men like him before—plenty of them—and began with the usual questions, his tone calm, practised, almost detached.

But the man's answer wasn't what he expected.

"Doc," he muttered, barely lifting his head, "it's not just the drink that's killing me. It's what's waiting for me. It's in my blood... in my family."

Lynch stopped writing, pen poised above the clipboard. His gaze sharpened. The man wasn't trying to excuse himself. He sounded matter of fact, defeated.

"What do you mean?" Lynch asked, quieter now, leaning in just a little.

"Cancer," the man said, voice flat, eyes on the ceiling. "Colon cancer. Took my parents. Took my uncles. Took my cousins. It's coming for me too, and I know it. That's why I drink. What else am I supposed to do—sit around waiting for it?"

The room was suddenly too still. Lynch didn't answer straight away. He just studied the man's face—lined with exhaustion, heavy with something deeper than illness. This wasn't panic or paranoia. It was certainty. And not the kind born of speculation. This was someone who had lived inside a legacy of death long enough to believe it was his inheritance.

"Tell me about your family," Lynch said after a pause, his voice low but pressing. "How many of them had cancer?"

The man let out a long breath, as if he'd been waiting for someone to ask. And then, name by name, he began to recount them. His father. His mother. Two uncles on one side. A cousin. Another cousin. Some had died young. Some not so young. But all to the same disease. Lynch listened, scribbling notes almost automatically, though his mind was already racing ahead. There was a pattern here—too defined, too recurrent to be chance. Something ran through this man's family like a thread, and no one had pulled at it yet.

After leaving the man's bedside, Lynch couldn't shake the thought. He had spent his career chasing answers, but this...this was different. It was personal now, grounded in a single, desperate human life. That night, as he reviewed his notes, he began to wonder how many families out there were living under the same shadow, waiting for cancer to strike like a thief in the night. He felt an urgency to act, a need to uncover the truth not just for this patient but for the countless others who might share his fate.

Lynch didn't waste time. Determined to follow the thread of this story, he set out to track the man's family. He bought a modest camper van and outfitted it with the bare essentials: a centrifuge, blood sampling kits, and a stack of notebooks to record family histories. On weekends, he drove through the rural landscapes of Nebraska, Kansas, and Missouri, following up leads and making quiet inquiries. He would sit in farmhouse kitchens with a cup of coffee warming his hands, asking gentle but deliberate questions.

"Mrs Miller," he'd say, pen poised, "tell me about your mother. Did she have any health problems? What about your siblings?"

The conversations were often raw, sometimes halting, but always disarmingly personal. For many families, cancer was no longer just a disease—it had become part of the fabric of their lives. It shaped how they thought about the future, how they raised their children, and how they mourned the past. But Lynch's presence—his sincerity, his patience—had a way of opening people up. He wasn't rushing through a checklist. He was listening, carefully, respectfully, as if every detail might matter. Because to him, it did.

With each visit, he pieced together the same story told in different voices: colon cancer appearing far too early, sometimes striking in the fourth decade of life or even sooner. Among the women, an unsettling number of cases of endometrial and ovarian cancer stood out—frequent, devastating, and eerily consistent. The pattern wasn't loose or vague. It was structured, persistent. The evidence was mounting, family by family, line by line.

But the work took its toll. Driving long hours across vast stretches of farmland, often alone, Lynch had plenty of time to sit with his doubts. What if he was wrong? What if what he was seeing was just coincidence—or worse, the projection of a theory he wanted too badly to be true? The medical community, at that time, had its own fixed ideas. Most cancer research was rooted firmly in environmental causes: carcinogens, diet, radiation, viruses. Genetics sat uncomfortably at the margins. To suggest that cancer could be inherited—that it might run through a family like an invisible thread—wasn't just controversial. It was, in some circles, considered reckless.

Some physicians warned that such ideas might lead patients to fatalism, to giving up on prevention or treatment. If cancer was inevitable, what was the point of trying to outrun it?

But Lynch couldn't ignore what he was finding. The stories weren't isolated. The patterns weren't rare. He saw the same configurations again and again, and they didn't look like chance.

He compiled his findings and submitted a research grant proposal to the *National Institutes of Health*. The hope was simple: funding would allow him to expand the study, reach more families, and begin to formalise what he had uncovered into a coherent scientific framework. But when the reply came, it landed like a blow. The proposal had been rejected—dismissed as too speculative, too ambitious, too far outside the current understanding of cancer.

Lynch sat with the letter for a long time. The words on the page echoed criticisms he'd heard before. Too much risk. Not enough data. It wasn't the first time he'd been told no, and it wouldn't be the last. But this time, the rejection cut deeper—not because it was personal, but because he believed he was holding something vital, something that could change the course of countless lives. And now the door had been shut in his face.

Still, he didn't stop.

Rejection was nothing new to him. He thought of his boyhood in New York, where survival had been its own kind of education, and of the battles he'd fought since—on ships, in hospitals, in rings, and in classrooms. This was just another one. Another place he wasn't meant to belong. Another truth others weren't ready to hear.

But Lynch wasn't doing this for recognition. He was doing it for the people whose stories he carried in his notebooks. For the families who had lived too long under a shadow no one had bothered to name.

Driving through the open plains of the Midwest, mile after mile, he understood exactly what was at stake. He was convinced—down to his bones—that recognising hereditary cancer could save lives. And he was determined to make the world see what he saw.

Nothing—not rejection, not scepticism—was going to stop him.

Henry Lynch's journey into the world of hereditary cancer was as much a personal mission as it was a professional pursuit. Armed with growing evidence, curiosity that bordered on obsession, and the kind of determination that can't be taught, he began building something that had never quite existed before.

He wasn't working alone.

One of his most important collaborators was a social worker named Anne Krush—a sharp mind and an even sharper sense of human need. Where Lynch brought the science, Krush brought something just as crucial: the ability to reach people on a truly personal level. She had an instinct for reading the emotional undercurrents in a room, for knowing when to push, and when to stay quiet. Her empathy didn't soften the work; it strengthened it.

Together, they formed the heart of a small but growing team, setting out across the Midwest in that same camper van, criss-crossing the backroads of Nebraska, Kansas, and Missouri. They arranged visits carefully, but the meetings themselves were anything but clinical.

Lynch and Krush would sit at kitchen tables surrounded by the ordinary chaos of family life—children's schoolbooks stacked on counters, dogs barking in the yard, a pot of coffee always brewing. These weren't interviews in the strict sense. They were conversations, often beginning slowly, winding through memories, grief, and unspoken fears. And always, the same question lingered just beneath the surface: was it happening again?

Lynch typically led with quiet confidence, his questions open but purposeful. "Tell me about your family," he'd say. "Who's had cancer? What kind, and how old were they?"

The responses were often steeped in a kind of weary inevitability. "Everyone on my mother's side had it," one woman told them, her voice flat with the weight of repetition. "It's just how it is in our family. We call it the family disease."

There was no melodrama in these accounts—just the steady accumulation of loss. Krush would listen with the same focus as Lynch, often catching the things that weren't said. After the formalities, she'd stay behind with family members who needed more time, more space. She understood that these weren't just medical cases. These were families haunted by a story that kept repeating itself.

While not medically trained, these families often had a profound awareness of their own medical histories. They referred to themselves as "cancer families," a term spoken with both inevitability and unity. Lynch was struck by how intuitively they grasped the genetic connections that bound their fates. Their knowledge came from funerals and hospital corridors, from conversations in kitchens and whispered warnings passed down through generations. Their words weren't just anecdotes to him; they were data, vital clues that would help piece together a larger picture. Lynch took careful notes.

In the evenings, he would sit under a desk lamp with a pencil and notepad, drawing out the family pedigrees by hand. Squares for men. Circles for women. Lines connecting generations. Shading to mark those who had died of cancer. What began as rough sketches soon evolved into detailed maps of inheritance—visual records of genetic devastation.

Again and again, the same motifs appeared. Colon cancer in fathers and sons. Endometrial and ovarian cancers in mothers and sisters. Often, multiple cases in a single generation. Often, striking before the age of 50. The precision of the pattern was impossible to ignore.

Alongside the medical records and autopsy reports, Krush and Lynch also gathered census data, family photographs, newspaper clippings—anything that could help confirm the history. The process was painstaking. The hours were long. But each new family added another piece to the puzzle.

Sometimes they organised informal reunions at local clinics or physicians' offices, inviting as many relatives as possible from a single family. These gatherings were equal parts medical investigation and communal reckoning. People would lean over the charts Lynch had drawn, pointing at boxes and circles, filling in missing names, correcting dates.

Laughter and tears mingled freely. Some joked about their odds— dark humour as a kind of shield—while others sat quietly, taking it all in. For some, it was the first time they'd seen their family history laid out in such stark, undeniable terms. What they'd long feared suddenly had a name, a shape, a pattern. And for many, that brought a strange kind of relief. At last, someone was paying attention. At last, someone was asking the right questions.

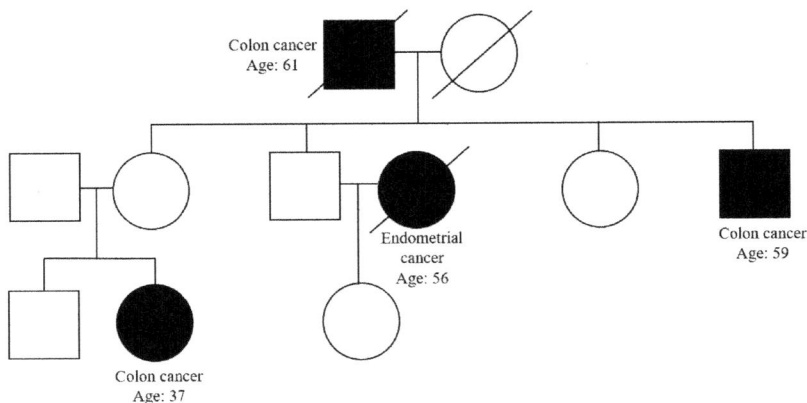

This pedigree is an example of Henry Lynch's mapping method, where squares represent males and circles represent females. Affected individuals are shaded and labelled with their specific cancer type, while deceased members have a diagonal line. The pedigree shows the autosomal dominant inheritance pattern, where cancer can be passed through either parent. Notably, not all carriers develop cancer, as seen with the mother of the 37-year-old female with colon cancer in this chart.

For years, the medical consensus held that when colon cancer appeared across generations, it could usually be traced to a condition called familial adenomatous polyposis—FAP for short. It was well-documented, familiar, and easy to spot. Patients with FAP developed hundreds, sometimes thousands, of polyps in the colon, usually starting in adolescence. The link to cancer was clear, the genetic mechanism increasingly understood.

At first, Lynch assumed this was the explanation for the families he was studying. It made sense. FAP was the only known hereditary syndrome associated with colorectal cancer, and these families had all the hallmarks of a genetic connection. But as he began combing through the clinical data, something didn't add up.

Many of his patients—far too many—didn't have polyps. Not in childhood, not in adulthood. No warning signs, no clusters of benign growths crowding the intestinal wall. Just cancer. Fast, aggressive, often fatal. They had the family history. They had the age profile. But they didn't have the defining feature of FAP.

It gnawed at him.

Night after night, Lynch sat with his hand-drawn pedigrees, flipping through medical records, cross-checking pathology reports. He retraced every case, every generation, looking for something he might have missed. But the answer kept staring back at him: this wasn't FAP. It couldn't be.

And if it wasn't FAP, then what was it?

He began to see what others hadn't: a distinct hereditary cancer syndrome hiding in plain sight. It followed rules of its own—fewer physical signs, a broader range of cancers, and an inheritance pattern that couldn't be ignored. Lynch didn't yet have a name for it, but he could see its silhouette forming in the data.

This realisation brought urgency. He had to share what he was seeing, even if it meant challenging accepted doctrine.

In 1964, he presented his findings at the annual meeting of the *American Society of Human Genetics.* He stood before a room of peers—some curious, others sceptical—and laid out the evidence: detailed family histories, early-onset colorectal cancer cases without polyps, and an emerging pattern that suggested something far more complex than FAP. His voice was steady, his charts precise. But the reaction was mixed.

Some attendees nodded along, intrigued. Others folded their arms, cautious, doubtful. It wasn't that they didn't believe Lynch's sincerity—it was that what he was proposing would require a shift in thinking. And medicine, like any institution, didn't shift easily.

Then, from the back of the room, came a voice.

Marjorie Shaw, a medical geneticist from the *University of Michigan*, approached Lynch after his talk. She had seen something similar, she told him—a family whose cancer history mirrored the pattern he'd described. The same early onset. The same absence of polyps. The same sense that something deeper was at work.

Their conversation that day marked the beginning of a quiet revolution.

Lynch returned to Nebraska newly energised. He and Shaw began collaborating, comparing notes, tracking down parallel cases. Over the following months, they worked with a small group of colleagues, including Anne Krush and Arthur L. Larsen, a pathologist at the *University of Nebraska Medical Centre.* Together, they began refining the criteria for what Lynch had started calling "cancer family syndrome."

In 1966, they published a report on two key families—Family N from Nebraska and Family M from Michigan. The paper was dense with detail, the language clinical, but the implications were seismic. The families had strikingly similar histories: multiple members with colorectal cancer, many before the age of fifty, and crucially, without the characteristic polyps of FAP. The term "cancer family syndrome" was now formalised, and though it would later be renamed hereditary non-polyposis colorectal cancer, it became known as *Lynch syndrome.*

But the publication was just the beginning.

Even with his findings in print and a formal name attached to the condition, Lynch knew there was still an uphill climb ahead. He hadn't set out simply to identify a syndrome—he wanted to change the way medicine understood cancer itself. That meant continuing the work, expanding the research, and proving beyond any reasonable doubt that what he was seeing in a few dozen families was not isolated. It was everywhere—hidden in plain view, quietly devastating generations.

So he widened his reach.

What began with a handful of families in the American Midwest soon grew into a much larger investigation. With the help of Anne Krush and a growing circle of collaborators, Lynch began tracing the genetic histories of families across the United States and eventually, across the world. He followed leads wherever they took him—from remote farming communities to bustling city suburbs, from small-town clinics to government health offices. Each family brought new names, new stories, new data.

He didn't rely on abstracts or second-hand reports. He went in person.

He traced the genetic histories of over 3,000 families worldwide, travelling to states across the U.S. and countries as far as New Zealand, Israel, Uruguay, and Europe. In every corner of the globe, the same shadow appeared: families losing members to colorectal and endometrial cancers in their forties, thirties—even twenties. Often, it was the same quiet realisation: *this has happened before.*

These weren't always easy visits. There were language barriers, cultural sensitivities, and sometimes, deep-rooted mistrust of institutions. But Lynch brought something institutions rarely did: presence. He showed up. He listened. He asked careful, respectful questions. And he explained what he was seeing—not as a detached scientist, but as someone who genuinely wanted to help.

Families opened their doors. They shared medical records, autopsy reports, and family trees drawn in pencil on yellowing notepads. Some produced boxes of old letters and death certificates, carefully stored in attics or sideboards. They brought him into their living rooms, introduced him to siblings and cousins, told him who had survived and who hadn't. For many, it was the first time they'd spoken openly about the pattern that had hovered over their lives for decades.

And in return, Lynch offered something that was almost radical in its simplicity: hope.

For much of his career, Henry Lynch had been working ahead of the curve—drawing conclusions the science couldn't yet fully support, trusting patterns that others couldn't yet see. He had spent decades building his case the hard way: by listening, by mapping, by showing up. But throughout that time, he knew the limits of observation. Pedigrees and histories could only take him so far. What he needed was molecular proof—evidence written directly into the genetic code.

That proof arrived in the 1980s.

The field of molecular genetics was entering a new era. Researchers had developed techniques that allowed them to examine DNA with unprecedented precision. Methods like microsatellite analysis allowed scientists to study specific DNA sequences, opening a window into the genome. For Lynch, it felt like the moment the language finally existed to describe what he had been seeing all along.

It began with a trickle of findings. Scientists at *Johns Hopkins*, *Yale*, and other research centres started identifying specific inherited mutations—changes in DNA that seemed to track through the same kinds of families Lynch had been studying for years. One gene in particular, *MSH2*, stood out. Then came others: *MLH1*, *MSH6*, *PMS2*. Each mutation corresponded with a loss of function in the body's ability to repair DNA errors—a glitch in the machinery that allowed cancer to take hold.

And there, in those sequences, was the confirmation Lynch had waited decades for.

The families weren't just passing down a predisposition. They were passing down a specific, identifiable defect—something you could see, test for, trace. The syndrome that had once been explained through grief and guesswork could now be located in a strand of genetic code.

It was, at last, irrefutable.

For Lynch, this wasn't a personal victory. It was a vindication of the people he'd worked for. The farmers and factory workers. The women who'd buried their sisters. The sons who had watched their fathers die before they reached middle age. They had always known something was wrong. Now the science had caught up to their stories.

As the data grew and laboratories confirmed what Lynch had long suspected, the tide began to turn. The medical community, once hesitant to embrace the concept of hereditary cancer, began to see its scale—and its implications. By the early 1990s, *Lynch syndrome* was recognised internationally as a distinct hereditary cancer syndrome, no longer a fringe theory but a cornerstone of medical genetics.

Around the same time, the discovery of the *BRCA1* and *BRCA2* genes—linked to hereditary breast and ovarian cancer—added momentum to the shift. What had once seemed like isolated family tragedies were now understood as part of a broader genetic landscape, one with predictive power, actionable risk, and life-saving consequences.

For Lynch, these advances weren't the culmination of his work— they were the foundation for what came next.

Now that the genetic signatures could be identified, he began urging the medical field to take the next step: prevention. This wasn't just about naming a syndrome. It was about what you could do with that knowledge. Screening, surveillance, prophylactic surgery— these weren't theoretical ideas. They were real-world tools, and Lynch was adamant that they be used.

In families with *Lynch syndrome*, he argued, colonoscopies should start not at 50, but in the 20s. Endometrial and ovarian cancers should be screened for early—and if necessary, prevented through surgery. The recommendations were bold, even controversial.

Some doctors weren't ready to suggest elective surgeries in the absence of disease. Others worried about the psychological toll of telling young adults they carried a cancer gene.

But Lynch didn't waver.

"In these families," he said, again and again, "preventive surgeries save lives,"

And they did.

One by one, patients who might have died in silence began surviving. Generations that had once braced themselves for cancer began to hope for something different. The arc of the story was changing—and Lynch was still there, at the centre of it.

Lynch's dedication to this mission was boundless. He began his days before sunrise, often by 3:30 a.m., fielding calls from patients and colleagues across time zones. He reviewed new research, updated patient files, planned the day's meetings, wrote letters, checked data. He never entirely moved away from pen and paper—he trusted the rhythm of handwriting, the feel of a pencil between his fingers.

Even as technology changed, Lynch kept close to the human heart of his work. He wasn't interested in becoming a figurehead, and he certainly wasn't interested in retiring. He still travelled. He still met families. He still lectured—often standing for hours, sometimes after barely sleeping. When a debilitating back injury left him bedridden in the 1980s, he didn't cancel a scheduled lecture. Instead, he was wheeled into the classroom on a hospital gurney, flat on his back, delivering the talk without missing a beat.

He was that kind of man.

Genetic testing for *Lynch syndrome* became standard in high-risk families. Guidelines were revised. Medical schools began teaching hereditary cancer syndromes as core content. His name—once a footnote, or a fringe curiosity—was now embedded in oncology textbooks.

More importantly, lives were being saved.

Early detection protocols were catching cancers years before they became lethal. Preventive surgeries were giving women the chance to live without fear. Genetic counselling was helping entire families understand their risks—and act on them. The fear that had once been passed silently from generation to generation was no longer inevitable. It was something that could be faced, planned for, even outmanoeuvred.

And Henry Lynch, even into his 80s, was still working. Still walking hospital corridors. Still lecturing. Still writing. Still listening.

When Henry Lynch died on the 2nd of June 2019, at the age of 91, it was in the same hospital where he had spent most of his career—*Bergan Mercy* in Omaha. It was fitting. That building had been his laboratory, his classroom, and his front line. He had walked its halls for decades, always looking forward, always asking what else could be done.

His life had taken him from the streets of New York, through the battlefields of World War II, into the boxing ring, and finally into the halls of medical science. Yet despite the many battles he had fought, his greatest victory was against the cancer that had claimed so many lives before he intervened.

What is Lynch syndrome?

Lynch syndrome is a hereditary condition that significantly increases the risk of developing certain types of cancer, most commonly colorectal cancer and endometrial (uterine) cancer. Unlike most cancers, which typically appear later in life, cancers associated with *Lynch syndrome* often develop at a younger age. This condition is caused by a mutation in one of the genes responsible for repairing DNA during cell division. Usually, these DNA repair genes act like a quality control team, fixing any mistakes that happen when cells copy themselves. However, when one of these genes doesn't work properly, errors accumulate over time, which can lead to cancer.

Because of the increased risk, people with *Lynch syndrome* are strongly advised to undergo regular cancer screenings, such as colonoscopies, starting much earlier than the general population. For some, preventive surgeries, like removing the colon or ovaries, may also be recommended to lower their risk of developing cancer. With proper monitoring and preventive care, individuals with *Lynch syndrome* have a much better chance of detecting cancer early, allowing them to lead healthier, longer lives.

CHAPTER 15

One of Us

THE SKY ABOVE Devon in the summer of 1846 was the kind of brilliant blue that seemed endless, arching over fields so green they looked freshly painted that morning. John walked a few paces behind his family, hands tucked into his pockets, eyes scanning the horizon as though it held the answer to an unspoken question. The breeze was soft, almost playful, carrying the sweet, earthy scent of wildflowers mingled with the faint brine of the distant sea. Out here, away from the cramped confines of his father's grocery shop, the air felt lighter somehow—easier to breathe.

For years, John had stood behind that counter, his hands dusty with flour or sticky with molasses, balancing weights on brass scales and tallying figures in neat, practised lines. Pulled from school at 14, the thought of a future beyond the shop seemed distant, if not impossible. His father had tried and failed three times to keep the business afloat. Each failure carved new lines into his father's face, and John felt the unspoken pressure to hold things together. His place was behind the counter, and he knew it. But still, there were flickering moments when he imagined a life outside those walls—some future shaped by more than necessity.

The trouble was, he didn't know what that life might look like. What did he even want? The question came and went like a shadow, leaving behind a familiar, frustrating ache. He pushed the thought aside for now and let himself enjoy the clean air and the way the sun caught the grass in a shimmer. For today, that was enough.

As they walked, the sky began to shift—the vibrant blue dimming under the weight of thickening clouds. A gust of wind tugged at John's sleeve, and he looked up just as the first heavy drops of rain began to fall. His mother's voice rose above the wind, urging them to hurry, her skirt flaring as she broke into a run. Ahead, nestled snugly between two ancient oaks, stood a small farmhouse, its chimney smoking faintly, promising shelter and warmth.

By the time they reached the door, the rain was falling intensely, drumming against the roof and splashing up from the ground in muddy bursts. The door creaked open to reveal a modest room—walls stained by years of use but softened by the glow of a crackling fire. A kind-faced family ushered them in, their voices warm and welcoming. John wiped the rain from his face and stepped into the room, his eyes adjusting to the firelight.

That was when he saw her.

She stood by the hearth, pouring tea into a row of mismatched cups. Her movements were unhurried, precise. Her head tilted slightly, as though she were listening to something no one else could hear. When she noticed him, she looked up—and for a moment, their eyes met. Then she returned to her task, unflustered.

John tried not to stare, but he couldn't help it. There was something about her—something that tugged at the edges of his attention and held it fast. Her round face had an openness, a quiet serenity untouched by the chatter around her. Her wide-set, upturned, almond-shaped eyes caught the firelight and held it. Her features were different—not startling but striking in a way that made him pause. She moved with deliberate grace, as though every action mattered. Even her voice, when she spoke, was soft and lilting, with a faint lisp that gave it a rhythm all its own.

John's chest tightened—not with pity, but with something deeper. Something unnamed. He had never seen anyone like her. She didn't laugh like the others, didn't chatter or smile in the same way. There was a difference, yes, but it wasn't diminishing—it was illuminating. He felt a pull he couldn't explain, as though something within him had been nudged awake.

Was she born this way? Did it matter? And why did he feel so drawn to her, as though recognising something he'd never known he was looking for?

The rain outside softened to a steady patter, and his family began preparing to leave. Just as John turned to follow them, the girl stepped forward. She moved with the same quiet presence, her gaze steady but gentle.

"Thank you for your help," he said, voice warm with genuine gratitude.

"You're welcome," she replied—slow, deliberate, but sure. Then, after a small pause, she asked, almost shyly, "What's your name?"

He smiled. "My name is John Down."

Weeks had passed, but the memory of that rainy afternoon in Devon lingered in Down's mind with surprising sharpness. He could still see the girl standing by the hearth, the way the firelight had caught the angles of her face, the quiet focus of her movements. Something in her expression, in the calm depth of her gaze, had marked him. It wasn't just curiosity she had stirred in him—but something else, something slower and more enduring. An ache, perhaps. A sense of responsibility that he couldn't explain or ignore.

He was only 18, but he felt it in his bones: he was meant for more than a lifetime behind a shop counter. He didn't yet know how or where that conviction would carry him, but he knew it had begun. The girl had lit a fuse.

Torpoint, his hometown, was not the sort of place where such aspirations were spoken aloud. It was a small, sturdy settlement on the Cornish side of the River Tamar, its cobbled streets filled with the sounds of blacksmiths at work, children shouting after market carts, and the din of the nearby naval base at Devonport. The docks clanged with the noise of hammering metal and shouted instructions, the air laced with the smell of brine, coal smoke, and damp rope. Life here was plainspoken and practical. Everyone had their place. And for John, that place had always been behind the counter of his father's grocery shop—where the air was thick with flour dust and the sickly-sweet residue of old sugar jars, and where the register's neat ledgers mapped the family's survival.

But the walls of the shop, once familiar and safe, had begun to feel smaller.

He had kept his ambitions to himself for years, not out of shame, but because he knew what the answer would be. Still, the words had been growing louder inside him, pressing at the back of his throat. That evening, as the shop shutters were drawn down and his father sat behind the counter tallying the day's takings, John knew it was time.

"Father," he said, his voice low but steady, "I want to study medicine."

The silence that followed was heavy, immediate. His father looked up slowly, his brow furrowing as if he hadn't heard correctly.

"Medicine?" The word landed like a stone.

He closed the ledger gently and leaned back in his chair, his shoulders hunched beneath the weight of years. His voice was rough, but beneath it lay something more fragile. Fear, maybe. Resignation.

"And what of this shop, John? What of your family? I've failed three times already to keep this place afloat. You've seen that. You know what it's cost us."

John said nothing, listening.

His father shook his head slowly. "I won't let it fail again. You're needed here. Your duty is to this family, not some foolish dream."

The words cut more than John expected, but he stood his ground. He had imagined this moment a hundred times—always with more eloquence, more courage. Now, the words came haltingly, but they came.

"I understand, Father. But I believe I can help people. Not just sell to them. I want to do more. I think I can."

His father didn't reply. He only nodded once, almost to himself, and went back to his ledger. The conversation was over. The decision had been made for him.

But John didn't let that be the end.

He found another way.

A few months later, he left Torpoint for the bustling, grimy sprawl of London's East End. It was a different world—louder, rougher, alive with both possibility and squalor. There, amid the soot and press of the city, he found an apprenticeship with a surgeon-practitioner. The work was gruelling, often bloody, and rarely glamorous. He spent his days tending to the city's poorest, learning by doing: pulling teeth, applying blisters, bleeding patients, compounding crude medicines. Outside, the streets reeked of horse manure, smoke, and damp stone, but within the noise and chaos, something in him was taking shape.

Every wound he bandaged, every patient he treated—even those who left without thanking him—brought him a step closer to something solid. The idea of becoming a doctor no longer felt like an abstract ambition. It was a path. And he was on it.

But as the months passed, he realised that practical work alone wouldn't be enough. If he wanted to call himself a physician— truly and properly—he would need formal education. So he took another step.

He applied to the *Royal Pharmaceutical Society*, enrolling in their rigorous examinations. It was a gamble. Money was tight, time even tighter, but he studied with the kind of quiet determination that didn't allow for failure. In just a year, he passed two major exams, earning recognition for his skill and sharpness. But even then, he knew. Pharmacy wasn't the destination, it was a rung on the ladder.

He returned briefly to Torpoint, not to stay, but to help. His new knowledge transformed the family shop. He began formulating remedies—proper remedies—that drew customers from across the region. Sales improved. The shop stabilised. And his father, now too frail to manage it alone, watched with quiet pride as John carried the business forward. There was peace in the house again. Hope, even.

When his father died in 1853, it came as both a sorrow and a strange, silent liberation. Grief, yes—but also freedom. The final thread tying him to the counter had been cut.

He didn't hesitate. He applied to the *Medical School of the London Hospital.*

Medical school in mid-19th-century London was not the polished institution it would later become. Inside, the London Hospital reeked of carbolic, blood, and wet wool. Lecture halls were crowded and poorly heated, and cadavers arrived in a state that could only politely be called inconsistent. But John Down thrived.

Money was tight, and the cost of education loomed large, but his sister and her husband welcomed him into their home, giving him the lifeline he needed. Living with them allowed him to dedicate himself wholly to his studies. Their home was warm, a place of respite surrounded by the pressure of academic life.

It was during one such afternoon at their home that Down met Mary Crellin, his sister's sister-in-law. The moment she walked into the room, he felt his breath hitch, the world narrowing to the warmth of her smile. Mary had a gentle and magnetic presence, and Down found himself drawn to her in a way that unsettled and thrilled him. What began as polite conversation soon grew into something deeper. Their romance was quiet but undeniable, a beacon of light amongst his studies.

Even as Down excelled, earning gold medals in medicine, surgery, and obstetrics, the memory of that rainy afternoon in Devon remained vivid, even years later. When his professors encouraged him to pursue a prestigious academic career, Down surprised them all. Instead of the expected path, he expressed a desire to work in a lunatic asylum.

"Why waste your talent there?" one of his mentors asked, genuinely baffled.

Down's answer was simple: "Because that's where I'm needed."

It was true. His work, he decided, would not be in the gilded halls of high society but among those society had cast aside. He would dedicate his life to those like the girl in the farmhouse.

The year was 1858, and John Langdon Down had barely settled into his new life as a young doctor when an unexpected opportunity presented itself. *The Royal Earlswood Asylum for Idiots*, tucked away in the quiet county of Surrey, was in urgent need of a new superintendent. The role promised a permanent salary, housing, and the kind of security Down longed for as he began to build a life with Mary. On paper, it was the kind of position that offered stability and status. But the decision weighed heavily. *Earlswood* was no ordinary institution. It was a place tainted by shame—a hidden world where society's most vulnerable were shut away, out of sight and out of mind, considered too broken, too inconvenient, to be part of public life.

When Down first arrived, the grim reality of the place struck him harder than he could have anticipated. The sprawling grounds, lush and green under the English sky, gave the illusion of tranquillity. But as he approached the thick stone walls, a heavy silence settled over him—broken only now and then by distant, piercing cries that seemed to bleed out from the building itself.

The smell met him as soon as he crossed the threshold: a sour, clinging mixture of damp stone, bodily decay, and untreated illness. The corridors stretched ahead, cold and dim, each footstep echoing sharply off bare walls. When he reached the wards, his breath caught. 15, sometimes 20 children crammed into a single, airless room—many sitting slumped against one another, others rocking silently in place. Their faces were pale, their eyes glassy and vacant. Some clasped their hands tightly in their laps, as if trying to tether themselves to the small, familiar movements that still made sense.

Down moved deeper into the building, each step peeling back another layer of heartbreak. In one corner, a boy, no older than 10, huddled alone, his clothes little more than rags clinging to his thin frame. Down crouched, tried to meet his gaze, but there was nothing there to meet—only a hollow, aching absence. In another room, an older resident bore the scars of years of violence: a twisted back, bruised skin, an expression so distant it seemed the spirit had long since retreated to some unreachable place.

That night, Down sat alone at the desk in his quarters, his head bowed in his hands. He had thought himself prepared. He had braced for hardship, for sadness. But he hadn't anticipated the sheer magnitude of neglect—the way suffering hung in the very air of the place, thick and choking. *Earlswood* wasn't a place of care; it was a warehouse, a human dumping ground where the "idiots" of Britain were abandoned, locked away, and forgotten.

The *Lunacy Commission*, charged with inspecting such institutions, had already condemned *Earlswood* in their reports. Mortality rates were appallingly high. Diseases like typhus and tuberculosis spread unchecked through the wards. The so-called treatments administered to residents veered closer to cruelty than medicine. The scandal had seeped into the public press, and *Earlswood's* name had become shorthand for institutional failure.

Down could have turned away. No one would have blamed him. But as he sat there, wrestling with the enormity of what he had seen, something else stirred within him. Not despair. Not revulsion.

Resolve.

He had no formal training in caring for the mentally disabled. He had no grand theories to lean on, no prestigious mentors in this field to guide him. What he had was simpler, and in its way, more powerful: a sense of compassion, an inquiring mind, and a quiet but unshakeable Christian conviction that every human life—no matter how fragile, how broken, or how misunderstood—had value.

He had taken this post for a reason. And he would not fail those who had no one else to fight for them.

The first step was hygiene. Down ordered every room to be scrubbed from floor to ceiling, the windows flung open to flood the dark corridors with fresh air and sunlight. Slowly, the oppressive gloom that had hung over *Earlswood* began to lift, replaced by a tentative brightness that softened the building's hard edges.

Next came the staff.

Down dismissed those who clung to the old ways of corporal punishment and cruelty, replacing them with individuals who shared his vision of compassion and respect. The whip was banished; in its place came patience, understanding, and the belief that dignity could heal where brutality had only wounded.

But Down knew that restoring humanity to the residents would require more than clean wards and gentler attendants. It was the small, almost invisible gestures that mattered most. He insisted that every resident, no matter their abilities, be taught to eat with a knife and fork. It wasn't about discipline—it was about demonstrating that they were worthy of civility, that they could aspire to normality. Children who struggled with bedwetting were no longer shamed; instead, they were gently woken during the night to relieve themselves, sparing them the daily humiliation they had once endured.

The changes were simple, almost unremarkable to an outside eye. But inside *Earlswood*, they rippled through the community like a quiet revolution.

Word of the transformation soon spread beyond the asylum's walls. The institution that had once been synonymous with failure and despair was now gaining recognition as a place of hope and renewal. The press, which had once condemned *Earlswood* as a national scandal, now praised it as a model of compassionate care. The corridors that had once echoed with cries and silence now carried the quiet hum of daily activity and sometimes, joy. The residents were healthier. Their eyes were brighter. Slowly, steadily, their spirits rekindled under the simple but radical knowledge that they mattered.

John Down and his wife, Mary, had done more than reform an institution. They had built a community—a living testament to what could be achieved when compassion replaced neglect.

Mary's role in that transformation was invaluable. With her background in the arts, she brought light and colour into *Earlswood's* once-bleak halls. She organised concerts and theatrical performances, introducing residents to the magic of storytelling and music. Soon the sound of a piano echoed through the corridors, and laughter spilled from rooms that had once been tomb silent. Mary also initiated art sessions where residents painted, crafted, and expressed themselves in ways many had never experienced before. With brushes and colours in their hands, some unlocked emotions they had never been able to share in words.

While Mary focused on creativity and emotional expression, Down concentrated on building practical skills that gave the residents purpose. He introduced training programmes in gardening, animal care, baking, and crafts. Residents learned how to tend vegetable plots, care for small livestock, and bake bread to supply the asylum's kitchen. Purpose replaced passivity. Contribution replaced isolation.

Physical health, too, became a priority. Down insisted on a balanced diet, ensuring that residents were nourished in body as well as spirit. Exercise became a regular part of daily life, tailored to each individual's abilities. He held a simple but powerful belief: that every resident, no matter how severe their disability, could live a productive and fulfilling life if given the right support.

Faith remained a cornerstone of his philosophy. Down believed that spiritual nourishment was just as vital as physical and intellectual growth. Morning and evening prayers became part of the daily routine, along with Sunday services where hymns filled the small chapel. Faith offered not only comfort, but dignity—a sense that even here, behind high walls, the residents belonged to something greater than themselves.

As word of *Earlswood's* transformation spread, doctors and social reformers from across Britain travelled to Surrey to see the changes firsthand. The asylum's once-empty waiting list swelled rapidly, as more families sought a place where their loved ones would be treated not as burdens, but as individuals. Down welcomed the visitors with open hands, eager to share the methods that had restored dignity to so many who had once been forgotten. Some left inspired, determined to take his ideas and seed them elsewhere, sparking quiet revolutions of their own.

But for John Down, it was never the accolades or the visitors that mattered most. His focus remained exactly where it had begun—on the residents themselves.

As *Earlswood* flourished, Down's curiosity deepened. The better he came to know those in his care, the more he began to notice something intriguing. Certain physical traits seemed to recur across many of the residents: round faces, almond-shaped eyes with an upward slant, flat nasal bridges, and short necks. These features stirred an old memory, one that had never fully left him. He thought of the girl by the hearth—the encounter that had first set him on this path.

253

The similarity between her features and those of the residents was too striking to ignore, and it sent his mind racing with questions.

What could these shared traits mean? Were they connected to their disabilities? Was this a condition he could one day define—or even name?

Around this time, Down encountered the writings of Johann Friedrich Blumenbach, the German scientist whose theories on the classification of human races were widely regarded within the Victorian scientific community. Blumenbach's system divided humanity into five broad groups: Caucasians (white), Mongolians (yellow), Ethiopians (black), Malays (brown), and Americans (red). He argued that each group had unique anatomical characteristics shaped by climate, geography, and history. Though Blumenbach's ideas were flawed and crude by modern standards, they reflected the prevailing intellectual framework of the age.

Down, always eager to expand his understanding, found himself drawn to Blumenbach's classifications. He was intrigued by the idea of systematically categorising the differences he observed—not to diminish individuality, but to comprehend it within a recognised scientific lens.

At *Earlswood*, Blumenbach's theories intersected with Down's daily observations. As he continued to study the residents, he noted something that fascinated him: many of the children shared physical traits that seemed to align with Blumenbach's descriptions of the "*Mongolian race*." In an effort to group and understand what he was seeing, Down began referring to these individuals as part of the "*Mongolian family*"—a term that, while now considered offensive, was at the time intended not as an insult, but as a means of bringing visibility to a population long ignored and misunderstood.

Determined to document his findings carefully and respectfully, Down turned to photography—a relatively new and powerful tool. But unlike the dehumanising "mug shots" typically used in asylums,

Down approached the camera with gentleness and dignity. He dressed each resident in their finest clothes, combed their hair, and posed them with care. His aim was not to showcase their differences as curiosities, but to capture their humanity—their poise, their individuality, their worth.

Over time, Down assembled a remarkable archive: more than 200 photographs, creating one of the most extensive collections of clinical photography in the Victorian era. But more importantly, these images offered something rare and precious—a compassionate window into lives that had too often been rendered invisible.

When his observations were complete, Down compiled his findings into a paper for the *London Hospital Reports.* Titled *"Ethnic Classifications of Idiots,"* the paper focused heavily on what he called the *"Mongolian family."* Down detailed the traits he had so carefully recorded: *"round faces"*, *"almond-shaped eyes"*, *"flat, pug-like noses"*, *"broad hands with short fingers"*, and later, the distinctive transverse palmar crease across the palm. He proposed, in line with the limited medical theories of the day, that parental tuberculosis might play a causal role—a hypothesis that would later be disproven, but one that reflected the scientific limitations of his time.

Today, his terminology and some of his conclusions are rightly seen through a critical lens. Yet Down's fundamental goal—to bring understanding, dignity, and recognition to a group society had long abandoned—was radical for its time. These were not *"idiots"* to him, nor objects of pity. They were individuals, worthy of care, worthy of study, and above all, worthy of a place in the world.

Photograph of a resident at the Royal Earlswood Asylum, taken by Dr John Langdon Down, c.1865.

Whilst John Down had always been the visionary force behind *Earlswood's* transformation, it was Mary who became its heart—a quiet, steadfast beacon for the residents. Tirelessly, she devoted her time and energy, often working late into the evening to enrich the lives of the children and adults who called the asylum home. Her gifts for music and art breathed new life into the once-muted halls, weaving colour and vitality into places where silence had long prevailed.

Mary orchestrated choirs, taught chants, and brought residents together in ways that transcended their disabilities. Music became a common language—a way to connect, to share, to celebrate. The joy radiating from the residents as they sang or performed was a kind of alchemy, a transformation that words could scarcely capture. Often, Down would stand quietly at the back of the room during these gatherings, his eyes misting as he watched Mary lead the residents in song, their voices rising in fragile, triumphant harmony. She believed, just as he did, that their lives were worth celebrating.

Down's love and admiration for Mary only deepened as he witnessed the transformation she brought to *Earlswood*. He saw how the residents lit up in her presence, how her warmth and encouragement coaxed out talents and confidence that had long lain dormant. He wanted her work to be recognised properly—not simply applauded in private, but acknowledged for what it was: vital, essential, equal to any physician's.

With that in mind, Down approached the board at *Earlswood* with a simple request: to place Mary formally on the payroll. It wasn't about the money. It was about respect. About recognising that the nurturing work she did—building confidence, sparking joy, restoring dignity—was no less important than the physical care administered by doctors and nurses.

The board's response was curt and dismissive. Women, they insisted, were expected to volunteer their time freely. Their labour was charity, not employment. Compensation was out of the question.

The refusal wounded them both.

Mary hid her disappointment with quiet grace, throwing herself even more passionately into her work with the residents. But Down felt the sting sharply. It wasn't only an insult to Mary—it was a stark signal that the institution he had poured himself into was unwilling to fully recognise the humanity it claimed to serve. A crack had opened, and though small at first, it was impossible to ignore.

Tensions simmered. When Down later requested funding to exhibit the residents' artwork at an international exhibition in Paris—an extraordinary opportunity to show the world the talents hidden within *Earlswood*—the board dismissed the idea outright. The creative achievements of the mentally disabled, they argued, were not worthy of such attention. It was a rejection that went deeper than budgets and policies. It was a rejection of his vision.

For Down, it was the final straw.

In 1868, with a heavy heart, he resigned from his position. He left behind the institution he had once reimagined, carrying with him a hard-won understanding: that true change required more than good intentions. It required the freedom to dream beyond what others could see.

Down was not ready to retreat into a quiet life as a physician in London. He had bigger plans.

Searching for a new beginning, he found it in a grand white mansion on Kings Road in Hampton Wick. The house stood proudly among the meandering hills and lush greenery, its sprawling grounds offering the perfect canvas for the vision he carried. The first time he stepped onto the property, he knew.

Here, he would create *Normansfield*—a sanctuary where mentally disabled children, especially those from privileged families who might otherwise have been hidden away, would not be discarded but educated, nurtured, and celebrated.

Down was particular in shaping his vision for *Normansfield*. Families who wished to send their children there had to meet a clear and unwavering condition: they could not simply leave them there to be forgotten. Down demanded their active involvement and lasting commitment to the residents' well-being. Education at *Normansfield* would be comprehensive—focused not just on survival, but on growth, learning, and personal fulfilment. Down wanted each child to develop the skills to live as independently and confidently as possible.

The setting itself seemed designed for renewal. The mansion's white walls gleamed under the sun, framed by vibrant gardens and wide lawns that were carefully maintained. Beyond the house stretched farms and stables, where residents tended animals and learned to care for living things. The daily rhythms of life unfolded in simple, grounding sounds: the soft clop of hooves on dirt paths, the clucking of chickens, the wind stirring the trees. For many children, it was their first experience of nature's quiet beauty. Gardening became a regular part of life, and Down encouraged the residents to take pride in growing vegetables and flowers. It wasn't just about teaching responsibility—it was about showing them their work had value. That they had value.

Inside *Normansfield*, the atmosphere was lively and purposeful. Workshops buzzed with activity as residents explored trades and crafts, from weaving to baking. One of the most beloved pursuits was puppet-making, a craft that became the heart of the small theatre Down had built on the grounds. But the theatre was more than just a stage for plays—it became a gathering place for concerts, performances, and Sunday Mass. Preparing for these events gave the residents a sense of excitement and pride, their hard work culminating in evenings filled with applause and celebration.

Down's belief in his residents' potential was quietly revolutionary. He observed early on that many of his "*Mongols*," as he still called them, had a remarkable aptitude for imitation. Where others might have dismissed this as mere mimicry, Down saw a gateway to learning. He grouped residents by ability, fostering small communities built on encouragement rather than competition. Together, they learned: weaving baskets, baking bread, tending gardens. With each new skill, their confidence grew—not just in their work, but in themselves.

Down's vision extended beyond the safe grounds of *Normansfield*. He took his residents into nearby towns, teaching them to navigate markets, to interact with vendors, to handle money. These small excursions mattered. Each one chipped away at the invisible walls society had built around them, replacing isolation with the possibility of belonging.

As word of *Normansfield's* success spread, the institution grew rapidly. By 1876, it housed over 100 residents; by the late 1890s, that number had swelled to 160. But the impact of Down's work reached far beyond the walls of his sanctuary. Slowly, he was changing how society viewed mental disabilities—not as shameful, hopeless conditions, but as part of the human experience.

Outside of *Normansfield*, Down remained deeply engaged with broader social issues. He became a vocal advocate for women's rights, pushing back firmly against the prevailing myths of his time. Many still believed that educating women would lead to an increase in mental disability among their children—a baseless, damaging claim that Down refuted with passion. Education, he argued, was a universal right, essential not just to individuals but to the health of society itself.

By the autumn of 1896, however, Down's once tireless energy had begun to wane. A lingering flu infection from six years earlier had weakened his heart and lungs, and although he continued to work with the same determination, his body could no longer match the demands he placed upon it. Pneumonia set in, and on 7th October 1896, at the age of 67, John Langdon Down passed away.

The news of his death rippled quickly through Hampton Wick. On the day of his funeral, black-veiled horses pulled the carriage slowly through the streets. Shops closed their doors in silent tribute. Curtains were drawn. People lined the roads, shoulder to shoulder, many weeping openly. They had not only lost a doctor—they had lost a man who had fought, tirelessly and quietly, for those the world preferred not to see.

At *Normansfield*, the mourning was even deeper. Residents and staff gathered in grief, many knowing instinctively what they owed to the man who had believed in them when few others would. Staff who had stood beside him through years of work and hope bowed their heads in silent tribute. For them, the loss was not abstract. It was personal.

Mary stood among them, silent, heartbroken. She had been more than his wife—she had been his co-visionary, his companion in every triumph and every setback. Together they had built something extraordinary, something that would outlast them both. And now, as she looked out over the faces of the residents they had nurtured, she knew: their sons would continue the work. Down's vision would endure.

For decades after John Down's death, the term *"mongolism"* continued to be used to describe the condition he had so carefully documented. His observations had been groundbreaking, but the language had aged poorly, becoming not only outdated but harmful. By the mid-20th century, scientists had discovered the true cause: an extra copy of chromosome 21—trisomy 21—unrelated to the *Mongolian race*. Researchers, particularly from Asia, voiced their objections to the outdated term, rightly recognising its inaccuracy and the harm it perpetuated.

In 1961, a group of geneticists published an open letter in *The Lancet*, urging the medical community to abandon *"mongolism"* in favour of a more appropriate term. Among them was a scientist who approached Down's grandson, still overseeing *Normansfield*, with a respectful request: could the Down family's name be used instead?

After careful consideration, the family agreed. To them, it was a fitting tribute—a way to honour a man who had dedicated his life to understanding, dignifying, and celebrating those the world had too often cast aside.

In 1965, the *World Health Organisation* officially adopted the term *Down's syndrome.*

It was a pivotal moment—not just for the medical community, but for John Langdon Down's legacy. His work had always been about more than diagnosis. It was about compassion. About dignity. About insisting, again and again, that every life mattered.

That those so often forgotten were, and always had been, one of us.

What is Down syndrome?

Down syndrome is a genetic condition caused by a person being born with an extra copy of chromosome 21. Typically, people have 46 chromosomes, but individuals with *Down syndrome* have 47. This extra chromosome affects their physical development and learning abilities, leading to characteristic features like a flatter facial profile, upward-slanting eyes, and a shorter height.

People with *Down syndrome* may take longer to reach developmental milestones, and their learning needs can vary widely from person to person. They are also more likely to experience specific health issues like heart conditions or hearing difficulties. However, with the proper support, healthcare, and education, people with *Down syndrome* can live happy, active, and fulfilling lives, contributing meaningfully to their communities and families.

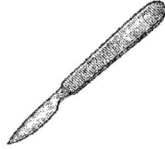

CHAPTER 16

The Battlefield
of the Brain

IN EARLY 1915, northern France groaned under the weight of a war that had turned fields into graveyards. What had begun as a clash of armies had sunk into a grinding, merciless stalemate. Trenches scarred the earth from the English Channel to the Swiss border, long, broken seams in a wounded land. Soldiers lived like spectres, trapped in a purgatory of mud, rats, and ceaseless artillery fire. The front lines barely shifted; every inch of ground gained came at a price too great to name. For those far from the fighting, the war was a slow, tragic spectacle. But for the men in the trenches, it was a daily reckoning with death—a crushing wave that ground down flesh and spirit alike.

Into this chaos stepped Harvey Cushing, a well-established American neurosurgeon. America had yet to join the fighting, but Cushing and a team of volunteers from *Harvard* had crossed the Atlantic, determined to offer what help they could. He was no stranger to pressure. For years, his steady hands and sharp mind had worked miracles in the quiet halls of Boston's hospitals. But this was something no training could truly prepare him for.

As their train rattled through the battered French country-side towards Paris, the scale of devastation unfurled outside his window. Where fields should have rippled green, only churned mud remained. Soldiers, half-crippled and blank-eyed, huddled in recovery stations along the tracks. Nurses moved among them like exhausted wraiths. At village platforms, young girls offered tricolour buttons to passers-by, their childish voices brittle with desperation, begging coins for the wounded.

Cushing sat rigid beside the window, his gaze locked on the passing ruin. Every mile seemed to strip another layer of innocence from the landscape. His chest tightened—not from fear exactly, but from something colder and harder: doubt. He rarely allowed himself that luxury. Outwardly, he remained composed, his face an impassive mask. Around him, his team murmured anxiously, their voices blurred to a distant hum.

One memory played over and over in his mind: the warning of an English nurse they had met along the journey.

"You haven't seen the true horror yet," she had whispered, her voice low and grim. "Men shot in the head, screaming, maddened with pain... bodies infected beyond all recognition. You're in for something dreadful."

The words lingered like a shadow, growing darker as the train rolled closer to Paris.

When they finally arrived, the *American Ambulance* was already a crucible of suffering, overflowing with the wounded. Their first patient came that very evening: a young French soldier, barely more than a boy, brought in with a bullet buried deep in his back. The injury wasn't remarkable by wartime standards. But the fear in the boy's eyes—wide, glistening, raw—left Cushing momentarily unsteady.

The operating room was a battlefield of its own: gurneys streaked with blood, the air thick with the sour-sweet smell of sweat, antiseptic, and open wounds. Life and death collided on every table. This was no longer the world of clean incisions and steady recoveries. This was survival surgery, and it was unforgiving.

Day after day, the tide of suffering grew. Twisted faces, shattered bones, wounds that festered before they could heal—each case another brutal reminder of what war did to flesh and bone. With every surgery, Cushing fought a silent war of his own, battling an enemy he couldn't outmanoeuvre.

One afternoon, bent over yet another disfigured patient, a French doctor leaned across and muttered something that struck him harder than any shellfire.

"You think this is bad?" the man said with a weary smile. "You haven't seen the real war yet."

Cushing stared at him, the words sinking deeper than he cared to admit. Wasn't this the real war? The mangled bodies? The haunted eyes? The endless procession of broken lives?

He had to see for himself.

One grey, bitter Sunday, Cushing left the relative shelter of the hospital and travelled towards the front. The countryside, once the pride of France, had been ground into a wasteland. Villages stood hollow and broken, their stone walls blackened by fire, roofs caved in like skulls. Trenches stretched across the landscape in endless, filthy lines, crowded with men who seemed less like soldiers and more like ghosts.

Artillery thundered on the horizon, and the sharp, jarring crack of rifle fire broke through the low mist. Even from a distance, the ground seemed to pulse with the violence of it. The closer Cushing came, the more the air itself seemed to tighten, as if the land were bracing against another blow.

He moved carefully along the battered trackways that threaded through the fields. Soldiers stumbled past him, mud caked to their uniforms, their faces slack with exhaustion. Near the edge of no man's land, he stopped. The earth here had been chewed to pieces. Shattered trees clawed at the sky. Bodies were dragged from the muck by stretcher-bearers, limp and ruined beyond easy recognition. Some moaned, low and broken. Others made no sound at all.

Cushing stood still, the cold biting through his coat, and watched.

There were no lectures or journals that could have prepared him for this.

War was not a matter of tactics and maps. It was this: men torn open, left to rot in frozen ditches; boys who would never see home again. Here, humanity itself seemed a fragile, flickering thing, struggling to stay alight against the vast, careless violence all around.

When he finally turned back towards the hospital, the images stayed with him—etched not in detail, but in feeling. He would never be able to look at his patients in quite the same way again. They weren't simply wounded. They were survivors of something far larger than bullets and shrapnel. Their bodies told only part of the story. The real wounds ran deeper, beyond the reach of any scalpel.

Cushing buried himself in work. He returned to the techniques he had spent years perfecting, relying on the rhythm of surgery to hold back the helplessness that gnawed at the edges of his mind. Still, the scale of the suffering seemed endless.

One afternoon, a French colleague beckoned him over to observe something new. A crude magnet had been fashioned to pull metal fragments from deep within a soldier's brain. Cushing was sceptical but intrigued. In the crowded operating room, with onlookers pressed shoulder to shoulder, he prepared a magnetised six-inch nail, hoping to extract a shard of steel.

The first attempt failed. Murmurs rippled through the room. He tried again. Still nothing. Doubt prickled at the back of his mind.

On the third attempt, the nail emerged—with it, a small piece of metal, glinting in the surgical light.

The crowd murmured in amazement, but Cushing barely registered it. His mind had already moved on. There was always another patient waiting. Another fragment of war to remove. Another life balanced on a knife's edge.

By May 1915, the strain had begun to show. Cushing, worn thin by the never-ending tide of injuries, arranged passage back to America.

The day of departure loomed, but he hesitated. Fatigue weighed on him, and the thought of leaving immediately felt insurmountable. Then, an alternative emerged: the *Lusitania*, a grand Cunard liner, was set to sail just a few days later. Larger, more comfortable, and promising a smoother journey, the ship tempted him. A few extra days of rest—surely, he reasoned, he deserved that much.

Little did he know, just days later, that very ship would be struck by a German U-boat's torpedo, sending 1,195 souls to the depths of the sea.

Two decades earlier, when Cushing was still a medical student, his hands trembled—a sensation that betrayed the steadiness he was supposed to embody.

He stared at the lifeless form of his patient, the silence in the room pressing down on him until he could hardly breathe. His heart hammered in his chest, each beat thick with guilt. Moments earlier, the man's pulse had been strong, steady under his fingers. He had administered ether, just as he had been taught, but something had gone wrong—terribly wrong.

Cushing had worked frantically, trying to will life back into the body before him, but the stillness was absolute. Final.

He remained frozen, unable to tear his gaze away from the young man's face. It didn't seem real. A mistake this final, this catastrophic, was something that happened to other people—less careful, less serious people. Not to him. And yet here it was, undeniable.

The room seemed to shrink around him, the operating theatre closing in like a trap.

He had barely begun his career, and already he had failed in the worst possible way.

Panic tore at him. How could he continue after this? How could he ever stand before another patient, knowing what had happened here?

His mind spiralled. And then, as so often before, his thoughts turned to his father.

William Cushing was a stern but steady presence in Harvey's life, a man whose expectations had shaped him from boyhood. Their family was steeped in medicine—his great-grandfather, grandfather, father, and elder brother had all carried the weight of that tradition before him. Joining their ranks had always seemed inevitable.

His father's voice echoed in his mind now, firm and unyielding. Be serious. Be disciplined. Avoid distractions. Baseball had been one such distraction—a passion that had once lifted him to captaincy at *Yale*. But now, in the hollow quiet of the operating room, Cushing wondered bitterly if all those hours on the field had stolen something he could never replace.

Would his patient have lived if he had been more focused? More devoted? Less drawn to games and glories that had nothing to do with the grim business of life and death?

The questions circled endlessly, their sharpness fading only as exhaustion took their place. Slowly, the panic ebbed, leaving behind a raw, unfamiliar stillness. He knew, even in that hollow moment, that he could not walk away.

If he abandoned medicine now, this failure would stain him forever. He would carry it like a weight he could never put down.

No. He would face it. He would learn. He would make sure it never happened again.

From that day, Harvey Cushing threw himself into his work with a focus that was almost ferocious. He scoured medical journals late into the night, sought out every mentor he could find, dissected every misstep until it yielded a lesson.

Mistakes would happen—that was the brutal truth of medicine—but he swore to himself that none would ever again be born of carelessness.

By 1895, as an intern at *Massachusetts General Hospital* (MGH), Cushing plunged into the demanding world of surgical practice. Interns lived within its walls, rarely sleeping, constantly on their feet, absorbing every detail of the craft. The operating room was a realm of speed and precision, where each second could mean the difference between life and death. Cushing thrived in the pressure, the brisk pulse of the theatre sharpening his skills. He didn't just observe; he participated. Every patient and incision was a lesson, a step toward mastering the art of surgery.

One of the most thrilling developments during his time at MGH was the arrival of X-ray technology. In late 1895, the news of Wilhelm Röntgen's discovery had swept through the medical world like wildfire. For centuries, surgeons had worked blind beneath the skin, relying only on touch, instinct, and educated guesswork. Now, for the first time, they could see inside the human body.

Cushing was captivated.

He wrote home to his mother, his excitement almost tumbling over itself: *"We won't be able to have secrets anymore,"* he marvelled, describing the ghostly images of bones captured on delicate film. It was as though a hidden world had been peeled back, revealing mysteries surgeons had only dared imagine.

But Cushing wasn't content simply to admire the new technology from a distance. Within months, he had helped secure X-ray equipment for the hospital's outpatient department, ensuring that MGH would be among the first in the country to harness its possibilities.

He immersed himself in the work, just as he did with everything else. He took radiographs himself, learned to mix chemical emulsions, and developed the films in darkened rooms heavy with the chemical stench of fixer and developer. He wasn't afraid to get his hands dirty—not when the reward was understanding.

In 1897, still barely beginning his formal career, Cushing published his first report: two cases of spinal gunshot wounds, their trajectories and locations identified through X-ray imaging.

It was a glimpse of what would become a defining pattern in his life: Cushing was never satisfied with the tools he was given. He wanted to push them further. To make them better. To find answers others had not yet thought to seek.

Yet, as exhilarating as these new frontiers were, something deeper had begun to take hold of his imagination: the brain.

At MGH, he had witnessed a handful of brain surgeries—rare, desperate operations attempted only when no other options remained. The risks were staggering. Even in the best of hands, most patients died on the table, or shortly after, overwhelmed by infection or swelling the surgeons had no way to control.

But to Cushing, the brain was more than a challenge. It was a frontier that mattered.

Unlike any other organ, it governed thought, movement, memory, the very essence of life. A body might survive a wounded limb. But a wounded mind? A wounded soul? That was something different. And so, quietly, almost instinctively, Harvey Cushing began to drift toward the one field of surgery that most others had the good sense—or fear—to avoid.

In 1896, Harvey Cushing moved to Baltimore to join *Johns Hopkins Hospital*, eager to work under the renowned surgeon William Halsted. It was a dream opportunity—*Hopkins* stood as a beacon of modern medicine, a place where innovation was not just encouraged but expected.

But for Cushing, the transition was jarring. He had come from the brisk, high-stakes environment of MGH, where every second in the operating room was a race against time, a battle against death itself. *Hopkins* was different. Here, precision reigned. Every incision was deliberate. Every stitch was placed with agonising care. The surgeries stretched on, slow and methodical, prioritising control above all else.

To Cushing, it felt dangerous. Every extra minute spent lingering over a wound seemed like an invitation to disaster. He couldn't help but remember the patient he had lost during anaesthesia—the life that had slipped away, he believed, because he hadn't moved fast enough.

The thought gnawed at him.

As the months wore on, his frustrations began to surface. At first, they were private: muttered complaints, impatient tapping of his foot during long operations. But soon, they spilled over into sharp words and open criticisms. He compared *Hopkins* unfavourably to MGH, where, as he liked to say, "one had to move quickly in order to keep warm."

The comments didn't go unnoticed. Among his colleagues, admiration for his skill was tempered by unease. His drive for speed and perfection seemed, at times, to verge on recklessness. To many, it felt like a refusal to respect the delicate work they were trying to master.

The tension eventually reached the ears of William Osler, the famed physician and one of the founding figures of *Johns Hopkins*. Osler had been watching Cushing for some time, recognising both the extraordinary talent and the growing restlessness. One afternoon, he found Cushing alone in a quiet corridor, hunched over a notebook, scrawling furious notes.

Osler approached with his usual calm. He folded his arms and spoke in a voice free of anger, but heavy with intent.

"Harvey," he began, "you have undeniable passion. No one doubts your skill. But surgery isn't just about speed. It's about respect—respect for the process, for your patients, and for the people you work along-side. You're not alone here, though sometimes it seems you forget that."

Cushing bristled at the words, his pride flaring, but he said nothing. Osler continued, his voice softening.

"I see potential in you," he said, "but I also see how easily this... impatience of yours could destroy it. This attitude, Harvey, would be absolutely fatal to your success here. The arrangement of the hospital staff is such that loyalty to one another, even in the smallest details, is essential. I know you won't mind hearing this from me, I have your interests at heart. But as the Psalmist says, *'Keep your mouth, as it were, with a bridle.'*"

The words hung in the air, heavy with truth. Cushing, used to excelling, wasn't accustomed to being chastised. His jaw tightened, the familiar sting of defensiveness rising in his chest. But Osler's concern cut deeper than any reprimand he had ever received. The older man wasn't trying to humiliate him; he was trying to save him...from himself.

Cushing nodded stiffly, muttering a polite acknowledgement, but the conversation lingered with him long after Osler had walked away. That evening, as he sat alone in his small quarters, he repeated the words repeatedly in his mind. For all his brilliance, he couldn't deny the truth in what Osler had said.

Instead of staying at *Johns Hopkins* to grapple with these lessons, he made a decision. Europe called to him, offering something that *Johns Hopkins* couldn't: a chance to reinvent himself on his own terms. There, he could seek out new knowledge, new techniques, and new ways to ensure no patient ever slipped away from him again. But deep down, he knew this journey wasn't just about technique or innovation. It was about finding a way to temper the fire inside him, to learn the patience and respect that Osler had spoken of.

Cushing's journey to Europe in 1900 marked a turning point in his career. From the moment he stepped onto the continent, Europe's energy invigorated him. The vibrant streets of Italy, the intellectual hum of Switzerland's university towns, and the rich history of France offered more than new knowledge; they gave him a fresh perspective. Each city brought new ideas, and with them, Cushing's curiosity deepened.

In Pavia, Italy, Cushing was introduced to a device that seemed almost too simple to be revolutionary: the Riva-Rocci pneumatic cuff. It was an unassuming blood pressure monitor with a rubber tube, a mercury column, and a hand pump. Standing in the quiet laboratory, Cushing watched as the needle rose and fell perfectly with each heartbeat. It was the first time he had seen blood pressure measured with such precision.

Cushing's mind raced. He immediately grasped its potential, envisioning how it could transform surgical safety. By measuring blood pressure alongside pulse and respiration, anaesthetists could monitor a patient's condition more accurately, providing surgeons with critical real-time data. It was a game-changer. This practice would soon become standard in operating rooms worldwide.

As he continued his journey through Europe, moving through Switzerland's university towns and the grand hospitals of France, Cushing's ambitions grew bolder. In Bern, under the guidance of Theodor Kocher and Hugo Kronecker, he began studying the brain's response to trauma in ways no one had systematically attempted before. He observed a phenomenon that puzzled even the senior physicians. When pressure inside the skull rose—whether from swelling, haemorrhage, or tumour—the body responded by raising systemic blood pressure, a desperate attempt to maintain blood flow to the starved brain.

This compensatory mechanism, now called *Cushing's reflex*, was elegant and terrifying. It revealed the lengths the body would go to protect its most vital organ.

Cushing's work didn't stop at observation. He was determined to turn theory into practice. He realised that blood pressure monitoring during surgery could do more than save lives under anaesthesia; it could provide invaluable insights into the brain's health during complex procedures.

In 1901, he published a groundbreaking report detailing the connection between elevated intracranial pressure and systemic blood pressure. Yet at the time, few appreciated the importance of what he was documenting. Some dismissed it as theoretical noise, irrelevant to the day-to-day demands of surgery. But Cushing knew better. The body was speaking, and he intended to listen.

Over time, his careful documentation and clinical applications proved undeniable. Once an experimental tool, the blood pressure cuff became indispensable in neurosurgery, marking a turning point in surgical safety and brain surgery.

Europe gave him more than knowledge. It gave him time to think, to question, to refine the restless urgency that had marked his early career into something sharper, steadier. When he returned to America, it was not as the same man who had left. He carried new tools, new questions, and, perhaps most importantly, a new sense of discipline.

This time, he negotiated the freedom to focus on what had become his true passion: neurosurgery.

Harvey Cushing's time at *Johns Hopkins* was a period of immense growth. Under the watchful eye of William Halsted, Cushing was pushed to his limits, not just to refine his surgical techniques but to embrace a level of precision that redefined the very nature of surgery. Halsted's methods were deliberate, almost maddeningly so. Every incision was calculated, and every stitch was placed with care.

To the untrained eye, it seemed unnecessarily slow. Still, to those who studied him closely, it was the foundation of mastery: gentle tissue handling, precise bleeding control, and an almost obsessive attention to detail.

In Halsted's operating room, perfection was non-negotiable.

The older surgeon rarely raised his voice or shouted orders, yet his presence demanded excellence. For a young, ambitious Cushing, the adjustment was difficult. He bristled under the constraints of Halsted's methods, but something clicked. It wasn't just about speed or skill; it was about control and respecting the fragility of the human body. Slowly, Halsted's lessons began to take root.

Yet Cushing's discipline came from more than Halsted's influence. It was rooted deep in his own past; in the habits he had shaped on the baseball fields of *Yale*. There, he had learned that even a moment's lapse in attention—a missed catch, a mistimed swing—could cost the game. That same ruthless focus now followed him into the operating room.

"Eyes on the ball," he would mutter under his breath during surgery, startling more than one distracted assistant. The lesson was clear: absolute attention, every second, every motion.

With each passing case, Cushing's skills sharpened, not just as a surgeon but as a problem solver. He sought out the most complex cases that others hesitated to take. At the time, operating on the brain was seen as a near-certain death sentence. But Cushing's meticulous nature paid off. His infection rates were strikingly low, and his patients began to survive in numbers that no one had thought possible. Brain surgery, once a last resort, began to offer a glimmer of hope.

Even as he transformed the field of brain surgery, Cushing's curiosity pulled him toward another mystery: the pituitary gland. This tiny, pea-shaped organ, hidden deep at the base of the brain, seemed to wield an influence far beyond its size. Its role in the body's internal regulation was still poorly understood, but Cushing was captivated. Here was an opportunity to explore a territory as complex and vital as the brain itself.

Between 1908 and 1912, he became consumed with unlocking its secrets. He encountered patients whose symptoms seemed almost fantastical—individuals with hands and faces enlarged grotesquely, suffering from a condition later understood as acromegaly, where the pituitary flooded the body with too much growth hormone. Others suffered from dwarfism, trapped in bodies that had been denied the very same hormones. Cushing saw these cases not merely as medical puzzles but as deeply human tragedies, and he was determined to understand them.

By 1912, after years of research and dozens of operations on patients with pituitary disorders, Cushing published *The Pituitary Body and Its Disorders*, a monograph that would change the course of endocrinology. In it, he documented cases of both hyperpituitarism—overproduction of hormones—and hypopituitarism, where the gland's output was perilously low. He linked clinical symptoms to anatomical findings with a clarity few had managed before. Among his many observations, he described a striking set of symptoms caused by a pituitary tumour: abnormal weight gain, muscle weakness, and high blood pressure.

The disorder would eventually bear his name—*Cushing's disease*—but at the time, it was simply another mystery he was determined to bring into the light.

Cushing's ability to link clinical observation with surgical intervention earned him international acclaim. But just as his career reached new heights, the world began unravelling. In August 1914, while on a rare fishing vacation in the quiet countryside, Cushing received news that war had broken out in Europe.

By March 1915, Cushing had assembled a team of 13 surgeons and 4 nurses from *Harvard*. They packed their supplies, said their goodbyes, and prepared to face a new kind of challenge that would test their skills, endurance, and humanity.

By May 1915, the war had left Harvey Cushing utterly exhausted. Months of tireless work at the front, operating on shattered bodies, witnessing unimaginable horrors, and facing a constant tide of suffering had drained him, body and soul. The brutality of the battlefield had seeped into his very bones. Though he had fought without pause to save lives, the destruction felt insurmountable. With all its chaos and senselessness, the war had begun to erode him. It was time to return home.

Cushing had arranged passage on the *Lusitania*, set to depart from England. It would be, he thought, a brief reprieve from the chaos, a chance to breathe after months of strain. Arriving in England, he felt a fleeting sense of relief. His first stop was Oxford, where he visited his old mentor and friend, William Osler. The Oslers had made England their home, and despite the heavy shadow of war, their warmth remained undiminished. Cushing was welcomed into their household with kindness, and for the first time in months, he allowed himself to relax.

The dinner that evening was a balm to his weary spirit. Conversation was deliberately steered away from the war, focusing instead on Osler's son, Revere. The boy impressed Cushing with his sharp mind and passion for rare books. Revere's curiosity and quiet confidence stirred something within him, a reminder of what was still worth preserving in a world being torn apart—the promise of youth, of potential untouched by war. For a moment, he could envision a bright future for the boy, filled with discovery and purpose.

The next morning, Cushing wandered the streets of London, savouring a rare moment of normalcy. The city, despite everything, bustled with life. Shopkeepers called to customers, and children laughed in the parks. As he browsed a small shop, the peaceful day was shattered by the sudden arrival of a policeman who burst into the room, his voice urgent and strained.

"They've got the *Lusitania*!" he cried.

Cushing's heart dropped. He hurried into the street, where the news was already spreading. Sandwich-board men carried signs splashed with headlines, and newspapers flew off the stands. The *Lusitania* had been torpedoed by a German U-boat before it could reach England. Over a 1000 passengers and crew were dead. Cushing stood frozen, his mind racing. Relief and guilt tangled within him. He had narrowly escaped, but so many others had not. The war's reach, he realised, extended far beyond the trenches. Even away from the front lines, no one was truly safe.

"When will England wake up?" he muttered under his breath, frustration and helplessness rising in the moment's stillness.

The following day, still shaken, Cushing boarded the *St Paul* for his return voyage to America. The mood aboard the ship was tense, almost sombre. Many of the passengers were Americans, their faces drawn with fear and anger. Life jackets were worn at all hours, and bitter conversations about the *Lusitania's* sinking filled the air. Curses were hurled at the Germans, but beneath the anger lay an unmistakable current of vulnerability.

On the first morning at sea, Cushing was in his cabin when a friend knocked urgently on his door, urging him to come to the deck. The *St Paul* had been passing through the wreckage of the *Lusitania* for over an hour. As Cushing stepped into the open air, the sight before him struck him to the core. The sea was littered with debris—overturned lifeboats, broken chairs, splintered wood. And then he saw them.

Bodies.

At first, it was a single figure, floating face down. Then, as the ship moved closer, more appeared. A woman and a child, drifting side by side, the child's small arm reaching toward the woman, as if seeking comfort even in death. The sight gripped Cushing, a wave of sorrow and anger surging through him. Around them, the wreckage stretched endlessly, scattered with more bodies—some tangled in lifeboats, others floating alone, lifeless and adrift. The cold indifference of the sea carried them along, silent witnesses to the war's senseless brutality.

Cushing stood rooted to the deck. The wreckage seemed to stretch for miles, but it was the woman and child that stayed with him, their faces seared into his memory.

As the *St Paul* sailed on, carrying its passengers toward safety, Cushing remained on deck, his thoughts adrift. Even the open sea, he realised, was no longer a refuge. Watching the last fragments of the wreckage disappear into the horizon, he knew the images would haunt him long after he reached home.

🩺

After his time at *Johns Hopkins*, Harvey Cushing's work at the *Peter Bent Brigham Hospital* in Boston focused heavily on brain surgery, particularly the formidable challenge of controlling haemorrhages. Brain operations were perilous, and bleeding remained one of the greatest obstacles. But Cushing's efforts led to transformative breakthroughs: he introduced adrenaline to reduce bleeding from the scalp and developed the now-famous *"Cushing clip"* to control bleeding vessels within the brain. These advancements would become standard practice in neurosurgery, dramatically improving patient outcomes.

Just as his surgical innovations were reaching new heights, however, his work was interrupted once again. America had entered the First World War.

In 1917, with the United States fully committed to the war effort, Cushing was called into service. Promoted to senior consultant in neurological surgery for the *American Expeditionary Forces* in Europe, he returned to the front lines, not just as a volunteer surgeon but as a key figure in the U.S. Army's medical division. As the war intensified, his responsibilities expanded. He was placed in charge of a base hospital, overseeing the care of soldiers suffering from devastating brain and nervous system injuries.

On the battlefields, Cushing encountered a new kind of injury: missile wounds that penetrated the skull and pulped the brain. These injuries were often fatal, but Cushing refused to surrender to the odds. He developed a new technique that saved countless lives: the use of a glass sucker to debride the missile tracks. The sucker, delicate yet powerful, allowed him to remove damaged brain tissue and embedded fragments like shrapnel while leaving the surrounding healthy tissue intact. With each procedure, Cushing thoroughly cleaned the wound tracks, closed the scalp, and moved on to the next patient, knowing that each life saved was a small victory against the overwhelming destruction.

The physical toll of this work, however, was immense. During the brutal Battle of Passchendaele, Cushing often operated for 16 hours a day, his hands remaining steady even as his body grew dangerously thin with exhaustion. The conditions were unforgiving—mud, cold, and the constant rumble of shellfire on the horizon. Yet his methods had a profound impact. Mortality rates for penetrating head wounds, initially staggeringly high, dropped significantly under his care—from 54.5% in the first month to 28.8% by the third. Despite the endless strain, Cushing knew his efforts mattered.

Throughout the chaos, he turned to his war diary as a way to process the overwhelming nature of his work. He wrote wherever he could—on the backs of X-ray reports, on scraps of temperature charts—recording not just the operations but the lessons learned, the lives saved, and the ones he couldn't reach in time. The sheer volume of his writing—over a million words—reflected his dedication to understanding and improving the treatment of war wounds.

But the intense conditions left their mark. Cushing developed peripheral neuritis, a condition that caused pain, weakness, and numbness in his limbs, symptoms that would trouble him for the rest of his life.

One night, as Cushing sat at his desk, exhausted from the day's operations, a letter arrived that set his heart racing. It was from Grace Osler, the wife of his dear friend William. Her words carried a mother's desperate plea. She wrote of her son, Revere, who was somewhere near the front lines at St. Julien, Belgium. Though she didn't say it outright, Cushing could feel the anguish behind her words, the fear that her son might already be wounded—or worse. She added that if the worst were to happen, they would find some comfort knowing he would be in the care of someone they trusted.

Cushing had met Revere two years earlier, a bright, promising young man, full of curiosity and potential. The thought of such a future being lost to the brutalities of war chilled him. Without hesitation, he penned a reply, asking for details about Revere's unit so he could try to locate him. He knew it was close to impossible to find one man among millions scattered across the trenches, but he couldn't stand by. The image of Revere—the boy he had once sat across from at dinner, now lost somewhere in the mud and blood— unsettled him greatly.

As he prepared to rest, his thoughts remained fixed on the Oslers. Then, a sharp knock broke the uneasy quiet. A messenger stood at the door, telegram in hand. The words hit him like a physical blow: *"Sir William Osler's son seriously wounded. Can Major Cushing come immediately?"*

Cushing wasted no time. In the pouring rain, he set off in an ambulance, the vehicle jolting and pitching over the rough, pitted roads as they raced through the Belgian countryside toward Poperinge. His heart pounded with a mix of urgency and dread. When he arrived, he was led quickly to Revere's side, and what he saw confirmed his worst fears. The young man lay still, his body battered by multiple wounds. One had torn through his upper abdomen, another pierced his chest just above the heart, and two more were lodged in his thigh. Cushing knew immediately that the odds were slim. There was only the faintest chance that Revere might survive.

Despite the dire situation, Cushing worked tirelessly. Opening Revere's abdomen, he found severe bleeding from the colon and surrounding vessels. His hands remained steady, performing each task with the care and urgency the situation demanded. But as the hours wore on, the truth became harder to deny. There was nothing more they could do. As dawn broke over the battered field hospital, Revere Osler—bright, intelligent Revere—slipped quietly away, another life claimed by the senseless brutality of war.

Cushing stood over Revere's lifeless body, an intense wave of devastation washing over him. He had seen death before—too many times—but this was different. Revere wasn't just another patient. He was a boy with a future, a son whose parents had trusted him to save. And now, that future was gone. He stared down at the still form, his heart breaking for the Oslers, who would soon receive the news no parent should ever have to bear.

As the war drew to a close, Harvey Cushing allowed himself a rare moment of relief. The endless days of bloodshed, the strain of operating under fire, and the unimaginable loss of life were finally behind him. For the first time in years, there was a sense of quiet. The British government honoured him with the *Companion of the Bath*, and the U.S. Army awarded him the *Distinguished Service Medal*. While humbling, these accolades paled in comparison to the quiet satis-faction he felt in knowing his work had saved countless soldiers. But what stayed with him most were the people—the connections forged amid chaos. Of all these, his friendship with William Osler left the most meaningful mark.

Cushing's admiration for Osler had never wavered. Osler's intel-lect, compassion, and dedication to medicine were qualities that Cushing carried into his own career. After Osler's death, Cushing knew the best way to honour his mentor was to capture his story. Writing Osler's biography became his way of preserving the life of the man who had shaped him so profoundly. When *The Life of Sir William Osler* was published, it was met with widespread acclaim, earning Cushing the *Pulitzer Prize* in 1926. But for Cushing, the recognition was secondary. The true reward lay in knowing he had done justice to Osler's memory.

Despite nearing 60, Cushing's drive for innovation never slowed. In 1926, he collaborated with Dr W. T. Bovie to develop a new method to control bleeding using high-frequency electrical currents. Ever the daring surgeon, Cushing was the first to test the technique in the operating room, successfully removing a large, highly vascular tumour with virtually no haemorrhage. The proce-dure was a triumph. Over his career, he performed more than 2,000 brain tumour operations, pioneering techniques that would shape the field of neurosurgery for generations.

In 1932, Cushing made another enduring contribution to medi-cine when he described what would later be known as *Cushing syndrome*. While preparing for a lecture on pituitary physiology, he noticed a distinct pattern among some of his patients: excessive weight gain, muscle weakness, and other physical characteristics. Initially, he suspected these cases might represent a variation of what he had earlier identified as *Cushing's disease*, caused by a pituitary tumour. But as he investigated further, he realised that not all these patients had pituitary tumours. Instead, many of their symptoms stemmed from overactivity in the adrenal cortex—the outer layer of the adrenal glands, located atop the kidneys. His analysis unified scattered observations made by others and provided a comprehensive understanding of the condition.

In his final years, Cushing faced his own health battles. His lifelong smoking habit had left him with vascular insufficiency in his legs, causing pain and difficulty walking. Though he resisted amputation, he eventually found relief after giving up tobacco—a decision he later lectured on, years before the medical community fully recognised the dangers of smoking.

Even as his health declined, Cushing never stopped contributing to the field. After retiring from *Harvard* in 1933, he continued teaching and research at *Yale*, where he served as the Sterling Professor of Neurology until 1937.

Harvey Cushing passed away in 1939 from complications of a heart attack. His death marked the end of an era, but his legacy as one of the greatest neurosurgeons in history remains.

What is Cushing syndrome?

Cushing syndrome occurs when the body is exposed to high levels of the hormone cortisol over a prolonged period. Cortisol, produced by the adrenal glands, is essential for regulating stress, metabolism, and the immune system. The pituitary gland, located at the base of the brain, plays a key role in this process by releasing a hormone called *ACTH*, which signals the adrenal glands to produce cortisol.

In *Cushing syndrome*, this balance is disrupted. This can happen for two main reasons: either from long-term use of steroid medications (commonly prescribed for conditions like asthma or arthritis) or from a tumour—usually in the pituitary gland, adrenal gland, or elsewhere in the body—that causes an overproduction of cortisol.

Symptoms of *Cushing syndrome* can vary but often include rapid weight gain, particularly around the face (sometimes called a "moon face") and abdomen, thinning skin that bruises easily, muscle weakness, high blood pressure, and mood changes such as irritability or depression. Over time, if left untreated, the condition can lead to serious health complications, including diabetes, heart disease, and brittle bones that are prone to fractures.

Treatment for *Cushing syndrome* depends on its cause. It may involve surgery to remove a tumour, medications to control cortisol production, or adjustments to the use of steroid medications.

CHAPTER 17

A Dubious Honour

SO FAR, WE have seen how many medical eponyms honour those who discovered them. Yet, it's easy to overlook the other person involved: the patient. Only a few diseases bear the names of those who suffered from them.

Carrión's disease

As the sun rose over the rugged peaks of the Andes, Peru faced a silent yet incessant menace. Long before the arrival of foreign settlers, the people of the highlands had known the painful marks of *verruga peruana*, or *Peruvian wart*—a peculiar skin eruption, unlike any other in the world, that lingered for weeks, sometimes months, on those it afflicted. In many cases, the nodules were preceded by periods of fever or general illness—common ailments that, at the time, drew little attention. The origins of the wart itself were shrouded in mystery, yet its presence was undeniable, woven into the fears and folklore of the Andean highlands.

But by the 1870s, a new and far deadlier spectre emerged: a fever that ravaged workers labouring on the ambitious trans-Andean railway. Sweating under the high-altitude sun, chipping away at the unforgiving rock, many succumbed to waves of intense fever and draining anaemia. The disease came to be called *Oroya fever*, after the mining town where so many fell ill and were buried in the shadow of the mountains.

Unlike other epidemics, it did not spread geographically but clung stubbornly to the Andean railway, striking anew every time the lines were repaired or rebuilt. For the Peruvian government, British railroad companies, and the U.S. Cerro de Pasco Copper Corporation, the disease was more than a health crisis—it was a menace haunting the lifeline of the nation's economy.

For Peru's doctors, however, it was a maddening enigma. Hidden behind symptoms that mirrored other illnesses, *Oroya fever* eluded diagnosis. Its victims died without warning, their bodies betraying no clear cause except for the insidious anaemia that drained the life from their veins.

One of those touched by the devastation was Daniel Alcides Carrión. Born in Cerro de Pasco, in the heart of Peru's mining country Carrión was no stranger to the suffering wrought by *Oroya fever* and *verruga peruana*. Tragedy came early: his father died when Daniel was just 8 years old, and the boy was sent to live with relatives in Tarma. But sorrow trailed him. In 1870, *Oroya fever* swept through Cerro de Pasco, taking nearly a third of the population. The losses carved deep into Carrión's heart, leaving him with a fierce, enduring need to understand the invisible enemy that had taken so much.

Gifted with a quick mind and an unwavering sense of purpose, Carrión rose academically, determined to follow a path in medicine. By his teenage years, he had made his choice clear: he would study the diseases that haunted his people. Yet the world around him offered little stability. Peru was caught in the chaos of the War of the Pacific, and every pursuit was made heavier by the weight of national suffering. Still, Carrión pressed forward, working as a medical practitioner during the conflict, honing his skills in Lima's hospitals. When peace returned, he turned back to the mystery that had haunted him since boyhood—the link between *Oroya fever* and the *Peruvian wart*.

In the corridors of *Dos de Mayo Hospital*, Carrión's mind churned with observations and questions. He had spent countless hours tending to patients from the highlands, watching some collapse under fevers and anaemia, while others bore the strange, reddened nodules of *verruga peruana*. To many, the two illnesses seemed unrelated—one deadly, the other disfiguring but survivable. But Carrión saw a pattern, an echo between the two, and he became convinced they were different faces of the same disease.

For him, this was no academic exercise. It was a personal crusade. If he could prove the connection, perhaps medicine could finally loosen the disease's grip on the Andes. Yet observation alone was not enough. Carrión needed proof that no one could deny.

So he made a decision as daring as it was reckless: he would prove it on his own body.

On the morning of 27th August 1885, Carrión stood in the *Sala de las Mercedes at Dos de Mayo Hospital*. The early light slanted through tall windows, throwing long pools of gold onto the whitewashed walls. Shadows stretched and shifted as if to mark the solemnity of the moment. At his side stood Dr Evaristo Chávez, who had agreed—reluctantly—to assist in the experiment. Chávez hesitated, every instinct urging caution. Carrión, however, was committed. His voice, when he spoke, carried the heavy certainty of someone who had already made peace with the risks.

"Whatever happens, it doesn't matter. I want to get inoculated."

Chávez, grim-faced, prepared the sample taken from Carmen Paredes, a young patient whose arms and legs bore the raised, blood-red lesions of *verruga peruana*. Carrión extended his arm. The room seemed to hold its breath. The needle slid into his skin, and in that instant, he crossed a boundary few would dare approach—becoming both subject and scientist, hunter and hunted. It was the ultimate test of his theory, his courage, and his willingness to risk everything for the sake of truth.

In the days that followed, Carrión carefully recorded his symptoms with the methodical eye of a physician and the urgency of a man whose life now hinged on each observation. At first, there was nothing—no fever, no discomfort, no sign that anything was amiss. For twenty-one days, he went about his work as usual, waiting, watching.

Then, on the 22nd day, the first cracks appeared. A deep, aching weariness seeped into his joints, followed by chills that no blanket could banish. His handwriting, once firm and clear, grew increasingly tremulous as he documented the fever's slow, creeping ascent.

"Until today, I had believed that I was alone in the invasion of the wart," he wrote, his pen faltering slightly. *"But now I am firmly convinced that I am attacked by the fever that killed our friend Orihuela."*

The realisation was both a triumph and a curse. Carrión had found the proof he sought: *Oroya fever* and the *Peruvian wart* were not separate diseases but two phases of the same sinister illness. Yet, as his body weakened, it became clear that the price of this discovery might be his own life.

With each passing day, Carrión's strength waned. The fever surged through him, sapping his vitality and leaving him pale and hollow. Severe anaemia took hold, draining the colour from his skin and the energy from his movements. Friends and colleagues, once inspired by his daring experiment, now looked on in horror as the disease consumed him. Despite their pleas for him to rest, Carrión remained resolute, determined to document every stage of the illness.

His journal became his anchor, the one thread he could still control. As the days wore on, his notes grew thinner, his sentences trailing off into ellipses of exhaustion. By the 26th day, delirium had begun to creep into his mind. His thoughts wavered; his words tangled. In rare moments of lucidity, he fought to hold the pen steady, to set down what he could before clarity slipped away again.

In one such moment, his trembling hand reached for his journal. He pressed it into the hands of a trusted friend, his voice no more than a dry whisper.

"I can't do it anymore," he murmured. "Please... continue for me."

His friends, torn between admiration and despair, took up the task. They watched helplessly as Carrión's body wasted away, the fever burning ever higher, the anaemia deepening to a deathly pallor. They documented it all: the sores that erupted on his skin, the hollowing of his cheeks, the swelling of his joints, the bloody sputum that stained his sheets.

By early October, Carrión's condition had worsened to the point of desperation. On 4th October, he was moved from *Dos de Mayo Hospital* to the French clinic, *La Maison de Santé*, in the heart of Lima—a final gamble for survival. There, in a white-walled room that smelled faintly of carbolic and despair, the doctors tried one last, desperate measure: intravenous injections of carbolic acid, a brutal and toxic intervention born more of hope than reason.

But it was too late. The disease had long since taken root, and no remedy, however harsh, could turn the tide.

As Carrión drifted in and out of consciousness, the fever gnawing at his mind and body, he summoned what little strength remained to speak to the friends gathered at his bedside. His voice, a frail echo of the man he had been, carried a glimmer of the same stubborn fire that had driven him to this point.

"I have not died yet, my friend," he whispered, a faint, almost mischievous smile playing on his cracked lips. "Now it is up to you to finish the work that has already begun, following the path that I have traced for you."

On 5th October 1885, 40 days after he had inoculated himself with the *Peruvian wart*, Daniel Alcides Carrión slipped into a final coma. His breathing grew shallow, his pulse thready. As the evening wore on, he surrendered quietly to death.

He was just 28 years old.

Yet his sacrifice was not in vain. Carrión's self-experimentation achieved what no theory or speculation could: it proved beyond doubt that *Oroya fever* and the *Peruvian wart* were manifestations of a single disease. His death, though devastating, was a clarion call to the medical world. What had once been a mystery now stood revealed through the bravery—and the blood—of a single man.

In the wake of his passing, Carrión's friends and colleagues refused to let his story fade into the dust of hospital records. They proclaimed him a hero of science, a martyr whose courage transcended the confines of his short life. Across Peru, his name became a symbol of sacrifice, not just for medicine but for the nation itself.

In time, 5th October—the day of his death—was declared the *Day of Peruvian Medicine*, a solemn tribute to the young doctor who had given everything in the service of knowledge. His memory became interwoven with Peru's national identity, an emblem of both scientific progress and human courage.

Today, *Carrión's disease*, carried by the tiny sandflies of the Andes, is recognised as a two-stage illness: an acute phase, known as *Oroya fever*, and a chronic phase marked by the nodular eruptions of *verruga peruana*. During the acute phase, patients suffer from high fevers, overwhelming fatigue, and devastating anaemia as the bacteria attack the red blood cells. Many workers on Peru's railways perished during this phase, their bodies unable to withstand the onslaught in the absence of effective treatments. Those who survived often entered the chronic phase, where clusters of reddish nodules would blossom across their skin, lingering for months or even years—a lasting reminder of the disease's silent cruelty.

Daniel Carrión's sacrifice solved a riddle that had tormented Peruvian doctors for decades. More than that, it galvanised a spirit of scientific inquiry that still burns in Peru's medical community today.

Lou Gehrig's disease

The stadium buzzed with anticipation, every seat filled as the crowd leaned forward, eyes glued to the field where Lou Gehrig made his entrance. His broad frame drew cheers, hands clapping, voices swelling in unison. He moved with a quiet, steady confidence that electrified the air, his modesty somehow amplifying the respect he commanded. As he stepped up to the plate, his stance perfectly balanced, hands wrapped firmly around the bat, every fan leaned in, sensing they were about to witness something rare and precious.

This was the Iron Horse, and they knew greatness lived in his every movement.

For Gehrig, these moments beneath the floodlights marked the culmination of a journey that had been anything but easy. Born in 1903 in East Harlem to German immigrant parents, his early years were shaped by hardship and perseverance. The Gehrigs lived in a modest apartment, a world away from the grandeur of the ball-parks that would one day echo his name. His mother, Christina, worked tirelessly as a laundress to keep the family afloat.

"We are not poor," she would tell Lou firmly, even when their pantry stood nearly bare. The conviction in her voice stayed with him, instilling a quiet dignity that would define him for the rest of his life.

The neighbourhood, however, was less forgiving. In the shadow of the First World War, prejudice against German Americans ran deep, and young Lou bore the brunt of it. Slurs like "dirty Hun" followed him through the streets of Washington Heights. He clenched his fists but never retaliated. His shoulders hunched, but his gaze remained steady, fixed on a future he had not yet seen but refused to surrender.

That same quiet resilience carried him to Columbia University, where a football scholarship cracked open doors that once seemed impossibly far away. Gehrig initially planned to study engineering, dutifully pursuing a stable, respectable profession. But it was baseball that ultimately captured his heart. A Yankees scout spotted his raw talent in 1923 and signed him, hurling Lou into a world that still felt distant and dreamlike to the boy from East Harlem.

Within a few short years, Gehrig's towering strength and natural grace earned him a place on the Yankees' roster. His swing—fluid, lethal—soon made him a star. But despite his stupendous rise, he remained grounded. He jogged across the field with his head slightly bowed, as if humbled by his own abilities.

As Gehrig steadied himself at the plate that evening, the crowd hushed. Anticipation hung heavy in the warm evening air. The pitcher wound up. The ball hurtled forward—and with a crack like a gunshot, Gehrig's bat sent it soaring deep into the stands. The crowd exploded, rising to their feet as the ball vanished into the night.

This was the Lou Gehrig they adored: a man who made greatness seem effortless.

His career soon became the stuff of legend. Over 17 seasons with the New York Yankees, Gehrig hit 493 home runs, became the first player to smash 23 grand slams, and played an unbroken streak of 2,130 consecutive games—a record so untouchable it would stand for decades. He led the Yankees to seven World Series victories and was named Most Valuable Player twice. When the Yankees retired his jersey, it was the first time in baseball history such an honour had been bestowed—a permanent tribute to a man who had become larger than the game itself.

Yet the numbers only told part of his story.

Lou's appeal ran deeper than statistics. His teammates spoke often of his reliability, his steady leadership, and the quiet way he carried both victory and defeat. He was no showman, no braggart. He was a craftsman, treating each game with respect, every at-bat as an opportunity to earn his place all over again. Fans admired him for what he could do on the field and the humility that accompanied his greatness.

As the 1937 baseball season wound down, Lou Gehrig, still dominating the field, began to cast an eye beyond the diamond. Encouraged by his business manager, Christy Walsh, he explored opportunities that stretched past baseball's chalk lines. Endorsements came first, then barnstorming tours across the country. And in an unexpected twist, Hollywood came calling.

The idea of the Iron Horse trading pinstripes for a cowboy hat seemed almost comical, but Lou embraced it with the same open-hearted enthusiasm he brought to everything else.

"I know I'm no actor," he said with a sheepish grin, "but I'm going to give 'em my best."

In early 1938, Lou and his wife, Eleanor, flew to California. Soon after, he found himself on the dusty set of *Rawhide*, a B Western packed with horse chases and saloon brawls, offering Lou a chance to step outside his comfort zone. Clad in cowboy boots and a ten-gallon hat, perched awkwardly in the saddle, he couldn't hide his delight.

"Boy," he said one afternoon between takes, flashing that unmistakable grin, "I've never had so much fun in my life."

The filmmakers, knowing better than to stretch Lou's dramatic talents too far, cleverly wove his athletic gifts into the story. In one memorable scene, his character, facing down a gang of outlaws, grabs billiard balls from a saloon table and begins hurling them with uncanny precision—knocking out villains one by one, each throw as sharp and devastating as one of his line drives.

It was absurd, charming, and somehow perfect. The Iron Horse, conquering the Wild West with nothing more than his throwing arm.

When *Rawhide* premiered, it received warm if modest praise. Critics were kind; audiences, charmed. Lou's performance had an earnestness that made up for any rough edges. For Gehrig, the whole adventure had been a joyful detour. Hollywood glittered, but it was never going to replace the field. By the end of the year, he was back where he belonged—on the diamond, bat in hand, ready for another season.

But as Gehrig returned to the diamond, something felt different—a subtle but undeniable shift. His once-unbreakable stride seemed just a fraction slower. At first, fans didn't notice, and Lou dismissed it as nothing more than fatigue from a long season. Yet, like a shadow creeping over the horizon, the first signs of something ominous had begun to emerge.

It began subtly—just a stumble, a slight hesitation as Lou Gehrig ran between bases, a fumbled ball that seemed unthinkable for someone of his calibre. For the fans, these minor slips could have been chalked up to fatigue or a rare lapse in concentration, but for Lou, they felt like cracks in the foundation. Each misstep carried a quiet unease, a gnawing sense that something wasn't right. The once fierce power in his swing faltered, his bat sending soft, harmless flies into the field where solid drives had once rocketed.

By the spring of 1939, the difference could no longer be dismissed. Routine plays eluded him. His sure-footed balance betrayed him, sending him awkwardly off benches and stumbling along baselines. In public, the Yankees dismissed the concerns as the natural toll of 17 brutal seasons. Privately, worried glances passed between teammates. Something deeper was at work.

James Kahn, a seasoned sports journalist, sensed what many dared not say aloud.

"I think there is something wrong with him," he wrote. *"Physically wrong, I mean. He is meeting the ball, time after time, and it isn't going anywhere."*

It wasn't just wear and tear. Something inside Lou Gehrig—something essential—was slipping beyond his grasp.

And he knew it.

Each stumble, each faltering swing, gnawed at him. His body, once a machine that never faltered, now seemed foreign and unreliable. He didn't speak of it—not to his teammates, not even to Eleanor—but the unease festered in the quiet hours, when the roar of the crowd had faded and only the doubts remained.

At night, he lay awake, replaying every misstep in his mind, searching for answers that would not come. Fatigue? Overtraining? A passing illness?

He clung to hope, stubbornly, as only Lou Gehrig could. But deep down, a colder possibility was beginning to take root.

Lou Gehrig sought answers quietly at first. He met with doctors in New York, who performed routine exams and shrugged off his concerns. "It's probably just exhaustion," they said. But Lou knew better. This wasn't just tiredness. It wasn't just age. It was something darker.

As the 1939 season wore on and the questions inside him grew louder, he made a decision. He would travel to Minnesota, to the *Mayo Clinic*—a last, private hope that someone there could explain what was happening to his body.

When Lou arrived at the clinic, Dr Harold Habein greeted him warmly. Yet even in that first handshake, Habein felt a chill of unease. Gehrig's grip—once firm, unyielding—was oddly soft. As they spoke, Habein noted the shuffle in his gait, the slight tremor in his hands, the way his once-proud shoulders sagged as though bearing some invisible burden.

Excusing himself, Habein sought out his colleague, Dr Charles Mayo. When he found him, his voice was low and grim.

"Good God," he said. "The boy's got amyotrophic lateral sclerosis."

The diagnosis was crushing. ALS—three clinical letters that

meant everything Lou had built, everything he was, would soon begin to slip away.

Sitting across from his doctors, Lou listened as they explained. Motor neurons—those vital messengers carrying signals from brain to muscle—were dying. His muscles would wither. His movements would falter. Eventually, even the simple act of drawing breath would become a battle he could not win.

There was no treatment. No cure. Only decline.

At first, the words barely registered. ALS sounded distant, technical—a term for a stranger's misfortune, not his own. But as Habein and Mayo spoke of weakness, of loss, of inexorable progression, the reality pressed closer, more suffocating with every word.

Lou sat in silence, absorbing the weight of the future laid bare before him.

He did not rage. He did not weep. That was not his way.

Inside, though, something vast and hollow opened up—a grief too raw for words. The body that had carried him so faithfully across fields and seasons was abandoning him. And there was no way back.

The news of Lou Gehrig's retirement sent shockwaves through the world of baseball. Fans who had known Lou as "The Iron Horse," the steadfast cornerstone of the New York Yankees, struggled to reconcile this image of unshakable strength with the devastating reality of his illness. As the Yankees prepared to honour him, quiet grief settled across the nation—not just for the team losing its heart, but for a sport that would never feel the same without him. Lou wasn't just a player; he was a symbol of endurance, humility, and the resilience of the human spirit.

The Yankees, knowing that no ordinary farewell would suffice, planned a tribute that would match the man himself—simple, heartfelt, unforgettable. They chose a date: 4th July 1939. A holiday of independence and hope, now forever tied to a more bittersweet kind of courage.

As the day approached, Yankee Stadium transformed. Red, white, and blue bunting draped the railings. Vendors moved through the aisles with their usual trays of peanuts and programmes, but their voices were softened by the occasion's solemnity. Almost 62,000 fans would fill the stands, their cheers muted, their hearts heavy.

The sun beat down on the field, casting long, golden shadows across the grass where Lou Gehrig had built his legend.

Inside the clubhouse, Lou prepared himself with the same quiet composure he had shown every day of his career. He dressed slowly, methodically, pulling on his uniform like a knight donning armour for a final battle.

When he emerged from the dugout, the stadium roared—not with the fevered ecstasy of a World Series win, but with something deeper, more reverent. A standing ovation that said what words could not: *We see you. We thank you. We love you.*

Lou stood at home plate, gazing out at the vast sea of faces. Teammates, rivals, fans, friends—they had all come to bear witness. Some wept openly. Others stood stiffly, swallowing hard against the lump in their throats.

When Lou stepped to the microphone, his voice, steady but laced with raw emotion, broke the silence.

"For the past two weeks," he began, his voice steady, "you have been reading about a bad break."

A pause. The crowd held its breath.

"Yet today," he said, his voice catching just slightly, "I consider myself the luckiest man on the face of the earth."

The words hung in the air, astonishing in their grace. This man—who had been handed a death sentence—stood before tens of thousands and called himself lucky.

As he spoke, Lou did not dwell on what he had lost. He spoke instead of gratitude—for his teammates, his family, the game itself. He thanked the groundskeepers, the ushers, the fans who had cheered him through countless seasons. He even found space for humour, joking warmly about his mother-in-law:

"When you have a wonderful mother-in-law who takes sides with you in squabbles with her own daughter—that's something."

The crowd laughed through their tears.

He spoke of his parents—the sacrifices they had made—and of Eleanor, his wife, "a tower of strength," who stood in the dugout blinking back her own sorrow.

As he neared the end, his voice grew softer but never wavered.

"I might have been given a bad break," he said, his gaze steady, his words clear. "But I've got an awful lot to live for."

When he finished, a moment of silence swept through the stadium. Then the crowd erupted, not in the usual cheers of a baseball game but in thunderous applause demonstrating their admiration, heart-break, and gratitude. Beneath the applause, countless tear-streaked faces gazed at the man who had taught them not just how to play the game but how to face life with courage. Lou Gehrig had given them one final gift: a lesson in grace.

In the two years that followed his diagnosis, ALS progressed with a mercilessness that mirrored the cruelty of the disease.

Each day brought a new loss—small at first, but undeniable. His steps grew slower, his once-powerful hands trembled when he tried to button a shirt. Tasks that had been second nature now demanded exhausting effort. Writing letters became an ordeal. Swinging a bat was unthinkable.

Yet Lou chronicled every step with the same quiet honesty that had defined his career. He poured his observations into letters to Dr Patrick O'Leary at the *Mayo Clinic*, documenting his body's betrayal in steady, unsentimental prose.

In January 1941, in one of his most vulnerable letters, Lou admitted:

> *"As for myself, it is getting a little more difficult each day. One cannot help but wonder how much further this thing can go."*

Even as he confessed his growing fears, he ended the letter with a promise—to *"hold on as long as possible."* The resilience that had made him a legend on the field remained stubbornly intact, even as his muscles withered away.

Through it all, Eleanor stood at his side.

She had been his quiet partner during the glory years, content to stay in the background while Lou's star ascended. Now, in the face of unimaginable sorrow, she became his shield, his advocate, his anchor.

"There was no hope in it anywhere along the line," she later recalled. "Just downhill going, every day a little more downhill."

She watched him fade, *"dying away by inches."* She would find him shaking his head, frustrated by his body's failures, only to gather himself moments later, summoning a smile that broke her heart even as it strengthened her resolve.

By mid-May 1941, Lou's breath had grown shallow, each movement slower, more laboured. Eleanor described him in those final days as *"a great clock winding down,"* his immense spirit trapped in a body that could no longer obey him. On 2nd June 1941, at their home in the Riverdale neighbourhood of the Bronx, Lou Gehrig slipped into unconsciousness. He passed away later that evening, just a few weeks shy of his 38th birthday.

Lou's battle with ALS, previously known as *Charcot's disease*, brought the illness into public consciousness in a way few others could have. In his honour, it became widely known as *Lou Gehrig's disease*, a name that symbolised courage in the face of devastating odds.

Eleanor, with the help of their executor, George Polack, later founded the *Eleanor and Lou Gehrig MDA/ALS Research Centre* at *Columbia University*. Through their work, Lou's legacy extended far beyond baseball, offering hope to countless families fighting the same battle.

Lou Gehrig's story reminds us that strength is not measured by home runs or unbroken streaks, but by the way a person carries themselves when all strength is stripped away.

Christmas disease

Stephen Christmas was born on a chilly February in 1947 in London, where the charm of his last name wasn't lost on his parents. They named their sons Stephen and Robin with a sense of humour, nodding to the warmth and nostalgia of the Christmas season.

Their father, Eric, an actor, had spent the war years touring with the *Royal Air Force Gang Show*, bringing laughter to Allied bases during one of the darkest chapters of history. But with the war over and acting opportunities scarce in post-war England, the Christmas family packed their belongings and emigrated to Toronto, Canada, searching for a new beginning. Stephen's mother, Patricia, found work as a medical librarian, while Eric secured a position with the *Canadian Broadcasting Corporation*. Together, they worked hard to create a stable life for their two boys.

It was in Toronto that Stephen's parents first noticed something unusual. At just 20 months old, Stephen began experiencing unexplained bleeding episodes. A small bump during play, a tumble on the carpet—ordinary toddler mishaps seemed to spiral into extraordinary problems. His parents watched helplessly as small scrapes bled for hours, refusing to stop.

One particularly troubling episode brought them to *Toronto's Hospital for Sick Children*, where doctors ran a series of tests determined to uncover the cause of the persistent bleeding. It wasn't long before they arrived at the diagnosis: Stephen had haemophilia.

For Stephen's parents, the diagnosis was as bewildering as it was devastating. Haemophilia was a rare and poorly understood disorder at the time. It meant Stephen's blood lacked the proteins necessary for clotting, turning minor injuries into potential crises. His parents were told he would need constant vigilance and regular blood transfusions to replace the clotting factors his body couldn't produce.

Patricia, with the habits of a librarian and the fierce love of a mother, learned everything she could about the disorder. She became Stephen's first and most tireless advocate, ensuring that no detail was overlooked.

Stephen's paediatrician, Dr Bernard Laski, quickly grew fascinated by the case. Not only was Stephen's haemophilia severe, there was no known family history. An anomaly that suggested something unusual beneath the surface. Seeking confirmation, Laski consulted Dr Louis Diamond, one of North America's foremost haematologists, who agreed: Stephen had classic haemophilia.

Yet something still didn't sit right.

Dr Laski's curiosity deepened with every strange test result. Eventually, he sent Stephen's blood samples to Dr Robert MacMillan at *Toronto General Hospital*, whose laboratory could perform a more specialised analysis.

Dr MacMillan conducted a "mixing test," which involved combining Stephen's blood with another haemophiliac to observe clotting behaviour. Typically, when two haemophiliac samples are mixed, the clotting time remains unchanged. But with Stephen's blood, something extraordinary happened: the clotting time improved. It was an anomaly that pointed to the presence of an unknown factor—a missing piece of the puzzle that seemed to hold the key to understanding Stephen's unique condition.

The discovery set off a ripple of excitement and inquiry. In 1952, when the Christmas family returned to England for a visit, Stephen was admitted to a hospital in London for further evaluation. His blood samples were sent to Oxford, where two of the most respected researchers in the field, Dr Rosemary Biggs and Dr R.G. Macfarlane, took on the case.

Biggs and Macfarlane had already made significant contributions to the study of blood disorders, but Stephen's case presented them with something entirely new.

The samples were carefully analysed, and what they found was groundbreaking. Stephen didn't just have haemophilia—he had a previously unrecognised variation of the disorder. The issue lay not with the classic clotting protein associated with haemophilia but with the absence of an entirely different protein that no one had identified before. In honour of Stephen's case, this protein was named the *"Christmas factor."* Today, it is known as Factor IX, and its discovery fundamentally changed the understanding of haemophilia, marking a turning point in the treatment and classification of blood disorders.

Clotting factors are proteins in the blood essential for stopping bleeding. When an injury occurs, these proteins work together in a cascade—like a series of falling dominoes—to form a clot. In people with haemophilia, one crucial domino is missing, meaning the sequence cannot be complete, and bleeding continues.

For Stephen, that missing domino was Factor IX. His unique condition became known as Haemophilia B (Unlike classic Haemophilia A, caused by a deficiency of Factor VIII), or *Christmas disease*, a rare disorder primarily affecting males. For Stephen, this meant that even a tiny bump could lead to dangerous, prolonged bleeding. Painful joint bleeds, muscle haemorrhages, and life-threatening internal bleeding became an ever-present risk.

Before modern treatment, haemophilia was often a death sentence. Bleeding episodes could erupt without warning. Traditional methods—bandages, pressure, even stitches—were powerless against the silent failure of clotting. For Stephen and others like him, blood transfusions became the only lifeline, replenishing the clotting factors his body could not produce.

But the relief was always temporary. Each fall, each injury, each simple accident meant another trip to hospital, another transfusion, another gamble with fate.

Life became a precarious balancing act, with every day shadowed by the possibility of sudden crisis.

Yet even amidst these dangers, advances in treatment were stirring. The years after Stephen's diagnosis brought a revolution. By the mid-1960s, scientists learned to extract and concentrate clotting factors from donated blood, making targeted treatment possible for the first time.

For Stephen, it was a transformative change. Concentrated Factor IX meant fewer hospitalisations, greater independence. He could even undergo minor surgeries—once an unthinkable risk—with a much lower chance of catastrophic bleeding.

Home therapy soon followed, allowing haemophiliacs to administer infusions themselves, without the constant disruption of emergency hospital visits. It gave them something precious: a semblance of normalcy, a sense of control over lives that had so often been dictated by fear.

As Stephen grew up, these treatments became part of his daily routine. They were his armour, his shield against the dangers that once stalked him so closely. Yet the battle was never easy. Painful joint bleeds still came. The spectre of serious injury never truly faded.

Whenever he received the *Christmas factor*, it was a bittersweet reminder of his condition. Despite the physical toll and the knowledge that his life expectancy was limited—most haemophiliacs in the 1960s were not expected to live beyond 21—Stephen carved out a life on his own terms. He found solace in photography and later worked as a taxi driver, embracing a lifestyle shaped but never defined by his condition.

But the 1980s brought a new and invisible threat. The blood products that sustained Stephen and thousands like him carried an unseen danger: HIV.

At the time, blood supplies were not screened for contamination. The miracle of clotting factor replacement became, for many, a silent killer.

By his late 30s, Stephen received devastating news: he had tested positive for HIV.

He kept the diagnosis to himself, unwilling to burden his mother, Patricia, who had already carried the weight of his lifelong illness. But during a visit to British Columbia, she stumbled across his AZT medication—the only drug then available to slow the virus's brutal march.

The discovery crushed her. The realisation struck with such force that she suffered a stroke the very next day.

For Stephen, the guilt was overwhelming. He blamed himself for her suffering, even as she reassured him with the resilience that had carried their family through every trial. Yet the shadow of that moment never truly lifted.

Despite everything, Stephen did not retreat into despair. In the face of mounting illness, he found new purpose.

He became a passionate advocate for those affected by contaminated blood. Working alongside other activists, Stephen helped bring public attention to the tragedy that had devastated the haemophilia community. He contributed his voice to the *Archival Study*, a crucial initiative that gathered evidence for a formal appeal to the Canadian government.

Because of efforts like his, the victims of the tainted blood scandal eventually received recognition, apologies, and compensation— hard-won acknowledgements that dignity still mattered, even when justice came too late.

Stephen also participated in a television documentary, telling his story not out of self-pity but with the determination to ensure others would not be forgotten.

By the early 1990s, however, Stephen's health had begun to falter further. In 1993, he was diagnosed with advanced melanoma—a malignant skin cancer he had ignored for two years, unwilling to endure more doctors, more interventions.

In December of that year, his body, battered by haemophilia, HIV, and cancer, could no longer fight.

He entered palliative care as his final illnesses reached their inevitable end.

The Toronto *Globe and Mail*, reporting on his death, described him as "*not only an inspirational symbol for haemophiliacs but also an activist who was critical in bringing the tainted-blood issue to national attention.*"

Even as his health failed, Stephen remained focused on helping others, leaving a legacy of courage and advocacy.

On 21st December 1993, just days before Christmas, Stephen passed away quietly at the age of 46. The timing felt almost poetic, fittingly leaving the world with an inspirational Christmas story.

CHAPTER 18

The Silent Dissenter

THE SPRING OF 1939 was lush and quiet in Germany, a deceptive calm before the storm of war. Fields blossomed in emerald and gold, rivers flowed, and cities charged with life under the shadow of an iron-fisted regime. Yet, beneath this apparent vibrancy, there was a different kind of silence, a chilling, deliberate stillness that settled over hospitals and asylums. Here, in the halls of the vulnerable, the first whispers of an unspeakable plan began to take root.

The program did not arrive with fanfare or proclamations but through a mundane bureaucratic decree issued on 18th August 1939. Doctors, midwives, and nurses were ordered to report any infant born with physical or mental disabilities. It seemed like another routine directive in a nation tightly controlled by the state, no different from the endless forms and protocols that governed daily life. Most complied without question, believing they were participating in harmless record-keeping. Few could imagine the cruel fate awaiting the children behind those neatly inked names and diagnoses.

By October, the deception deepened. Letters began arriving in homes already burdened by worry, their tone cordial, almost hopeful. They invited parents to admit their disabled children to newly established state-run paediatric clinics, promising specialised care and the latest treatments. For mothers and fathers clinging desperately to hope, these invitations must have felt like salvation.

But the promises were a lie. These clinics were not places of healing; they were killing wards. By year's end, infants and toddlers were dying—not of their illnesses, but from calculated overdoses of sedatives, or slow, clinical starvation. Their small lives were extinguished behind closed doors, in rooms scrubbed to a sterile, inhuman silence.

And it did not end there. The machinery of death, once set in motion, only expanded. In September 1939, under the pretext of wartime necessity, Adolf Hitler signed a secret order authorising the programme's extension to adults. The Führer Chancellery, a shadowy office operating beyond traditional government structures, became the engine of systematic murder. The programme was given a deceptively simple name: "T4," after its headquarters at Tiergartenstrasse 4 in Berlin.

Across the country, new forms began to arrive at mental institutions and care homes, masquerading as instruments of statistical research. They reduced human beings to a series of grim checkboxes: Could they work? How long had they been institutionalised? Did they suffer from epilepsy, schizophrenia, or other conditions? Behind closed doors, panels of respected physicians sorted these lives with chilling efficiency. Those marked as "unfit" were removed from their institutions and loaded onto buses or trains, bound for one of six discreet gassing centres. None would return.

The silence, inevitable and brittle, did not hold for long. In small towns and rural villages, grieving families began to compare notes. Too many death certificates bore the same vague causes: pneumonia, heart failure, a litany of fabricated ailments. Whispers thickened into rumours, and rumours flared into outrage. Among the most fearless was Bishop Clemens August von Galen of Münster. From the pulpit, his fiery sermons denounced the killings with a boldness few dared to emulate. His voice cracked the facade of secrecy, forcing the horrors of the programme into the open.

By August 1941, the backlash had grown too loud for even the regime to ignore. Hitler officially ordered the suspension of the T4 programme. But the killings did not stop. They merely shifted underground, adopting quieter, less conspicuous methods: starvation, lethal injection, deliberate neglect. The scope of victims widened too. The elderly, foreign labourers and bombing victims joined the ranks of the exterminated. By the war's end, at least 250,000 lives had been extinguished. It was a grim prelude, a dark foreshadowing of the Holocaust yet to come.

Among the victims was Helmut F., an ambitious young sailor whose path reflected the brutal efficiency and moral void of the euthanasia program.

Helmut was full of life. Friends described him as extroverted, even boastful. In 1938, at just 18 years old, he enlisted in the navy. Whether he was driven by a hunger for adventure or a desire to prove himself in a country demanding loyalty, no one could say for sure.

But his confidence, perhaps veering too often into impulsive recklessness, soon landed him in trouble. He bragged about missions, wore decorations he had not earned, and stole from his own unit. By July 1940, Helmut was imprisoned for unauthorised use of decorations, theft, and worst of all in Nazi Germany; treason, for sharing military secrets.

His troubles didn't end there. Imprisonment only magnified the traits that unsettled his superiors: restless energy, compulsive lying, flashes of grandiosity, and an almost reckless indifference to rules. Diagnosed with schizophrenia at the Navy Hospital in Kiel, he was transferred to the *State Mental Hospital* in Zschadrass. There, the diagnosis shifted: Helmut wasn't schizophrenic but displayed "psychopathic personality," they wrote, noting no genuine psychosis, merely a deep-seated disregard for authority and consequence.

Under the Nazi regime, that distinction could mean life or death. A diagnosis of schizophrenia might put a man on the list for "incurable" patients—euthanasia. "Psychopathic," by contrast, implied moral defect, not mental incapacity; the state could still hold him fully responsible, punishable, and—if useful—retain him for labour or service.

By the time Helmut was returned to Kiel, his fate had been all but sealed. His so-called "behavioural abnormalities" led to another transfer, this time to the *Psychiatric Clinic of Kiel University*. In March 1941, a naval court, relying on the testimony of Senior Assistant Physician Dr Elste, swung the pendulum back to schizophrenia.

Whether Dr Elste had overlooked the earlier diagnosis or simply ignored it remains an open question. What was clear was this: Helmut had entered a system where medical labels were not diagnoses but death sentences. His life now hung by the thinnest of threads.

It was in this perilous state that he came under the care of the *Psychiatric Clinic* at Kiel—and into the orbit of one man who might yet determine whether he lived or died: a consulting psychiatrist named Dr Hans Gerhard Creutzfeldt.

In 1885, along the quiet waters of the Süderelbe in Harburg, Hans Gerhard Creutzfeldt was born into a life of comfort and purpose. As the eldest son in a family steeped in professional tradition, he grew up under the steady influence of his father, Otto, a dedicated physician and public health officer. From an early age, Otto's commitment to healing and the welfare of others shaped the boy's imagination. Medicine was not simply a career; it was a calling, a fusion of intellectual integrity and deep humanity.

Recognising his son's potential, Creutzfeldt's parents made sacrifices to ensure he received the best education possible. After two years in a local school, he was sent to the prestigious *Johanneum* in Hamburg. There, immersed in a classical curriculum rich in history, philosophy, and the arts, he thrived. The humanities sharpened his mind and deepened his sense of reflection—traits that would later define both his medical practice and his moral compass. By Easter 1903, diploma in hand and a keen curiosity for the human body alive within him, Hans Gerhard Creutzfeldt enrolled at the *University of Jena* to study medicine.

From the outset, his academic path was marked by distinction. By 1905, he had passed his first medical examinations with "High Distinction." Seeking broader horizons, he transferred to Kiel's *Christian Albrecht University*, where he completed his state exams with honours in 1908. Yet it was during his practical training at Hamburg's *St George's General Hospital* that a new fascination took hold—a fascination with the mysteries of the human nervous system.

Under the mentorship of the pioneering pathologist Morris Simmonds, Creutzfeldt embarked on his first major research project. His dissertation, a study of the pituitary gland, earned him his medical doctorate in July 1909. Even at this early stage, it was clear that his ambitions reached beyond the lecture halls and clinics of Germany.

In August of that same year, driven by a hunger to see the world, he volunteered for the German Imperial Navy. As Assistant Naval Surgeon of the Reserve, he found a new rhythm aboard merchant marine ships. For the next year and a half, he sailed across oceans.— To the teeming ports of East Asia, the sunlit shores of Australia, and the vast cities of America. These voyages left a lasting imprint. Beyond their medical lessons, they exposed him to the breadth of human experience and strengthened his love for literature and philosophy. With a small library of Greek and Latin classics tucked into his quarters, he spent long nights wrestling with the great questions of existence beneath foreign skies.

By 1912, the pull of discovery drew him back to Germany and the frontiers of medical research. In Frankfurt am Main, he immersed himself in the microscopic structures of the brain—nerve cells, fibres, and glial networks—seeking the hidden architecture of thought, emotion, and memory. His growing expertise soon opened doors, and in March 1913, he joined the *Royal Psychiatric and Neurological Clinic* at the *Silesian Friedrich Wilhelm University* in Breslau.

The position, unpaid though it was, offered something far more precious: the chance to work under Alois Alzheimer himself.

For Creutzfeldt, Alzheimer was not merely a mentor but a revelation. Alzheimer's groundbreaking work, linking clinical symptoms to structural brain changes, resonated deeply. Neuropathology, under his guidance, was no mere theoretical pursuit—it was a map to understanding human suffering at its root. In Alzheimer's lab, Creutzfeldt found not just intellectual stimulation, but a cause to which he would dedicate his life.

Then came Bertha Elschker.

Born on 8th December 1890, the youngest of five children—two already institutionalised for intellectual disability—Bertha spent most of her childhood without obvious neurological disease. In 1899 she was placed in a convent school, more for supervision than for medical care. By her late teens, however, warning signs surfaced: a stiff, awkward gait noted in 1910, along with attacks of tremor and dramatic "hysterical" spasms.

Her condition worsened in the spring of 1913. Her walk grew unsteady, her speech wandered into fragments, and bouts of unprovoked laughter or rage burst through her previously quiet demeanour. When the 22 year old arrived at Alzheimer's neurology clinic in Breslau that June, she was scarcely more than a shadow of herself: spastic, incoherent, her movements jerky and unpredictable, her speech reduced to echoes and disjointed ramblings.

She mimicked the words of others in a singsong parody of conversation. Her trembling limbs betrayed a nervous system in collapse; her staggering gait hinted at a deeper, more insidious ruin. Yet perhaps most heartbreaking were her wild emotional shifts, sudden bursts of laughter or rage that seemed to come from nowhere, hinting at the internal chaos consuming her.

Creutzfeldt was struck by her plight. In Bertha, he saw not merely a case study, but a young woman being devoured by forces no one yet fully understood. He approached her illness with compassion, documenting every tremor, every whispered fragment of thought, as though by sheer thoroughness he might somehow rescue her from oblivion.

Even Alzheimer himself was drawn into the mystery. Together, mentor and protégé pored over Bertha's deteriorating condition, searching for patterns, for meaning hidden in the wreckage of her mind. They could slow her decline, perhaps ease her symptoms, but they could not save her.

Bertha died just two months later, in August 1913. Her death, though genuinely sorrowful, offered one last chance: the post-mortem examination. Under Alzheimer's supervision, Creutzfeldt undertook the dissection with solemn reverence. Somewhere within the folds of her ruined brain, he hoped to find the answers that had eluded them in life—a glimpse into the nature of the disease that had stolen her so cruelly.

He carefully preserved samples, filled notebooks with sketches and observations, determined to honour her brief, broken life with understanding. He believed that her suffering could serve a purpose, that through her, science might yet learn to save others.

But history had other plans.

The summer of 1914 brought a new cataclysm, one that would tear Creutzfeldt from his microscope and from the orderly pursuit of knowledge. The outbreak of war swept across Europe, and soon, the research that had consumed him was abandoned. Creutzfeldt was drafted into the German Navy, assigned to the front lines.

The quiet comfort of the laboratory gave way to the deafening chaos of battle.

Hans Gerhard Creutzfeldt was thrust aboard a naval vessel tasked with breaching the English blockade—a desperate mission on dangerous waters. In 1915, disaster struck. His ship was torpedoed.

The chaos was immediate and deafening: men shouting, metal screaming as it tore apart, the sea rushing in with unstoppable force. Creutzfeldt was hurled into the icy water, the cold cutting through him. All around, the wreckage of ship and crew bobbed helplessly. He fought to stay alive, clinging to debris as darkness and exhaustion pulled at him. Hours passed, or perhaps only moments—time became meaningless in the black, heaving sea.

Against all odds, he survived. Of the entire crew, only he and one other man lived to see rescue.

Relief was fleeting. Captured by the English, Creutzfeldt spent three long months as a prisoner of war. The days blurred into one another—a grey expanse of waiting, of thinking, of remembering. His mind returned again and again to the life he had left behind: the research cut short, the faces of patients whose fates he would never know. Most painful of all was the news that reached him from afar. In December 1915, Alois Alzheimer, his mentor and guiding light, had died.

The loss hit him like a blow. Though miles away, Creutzfeldt felt the light that had once illuminated his path gutter and fade. The future—his future—now stretched before him uncertain and dim.

Released from captivity in 1916, he was returned to active duty, this time as a troop doctor in a torpedo boat squadron. The war dragged on, and Creutzfeldt endured, carrying with him the private griefs and silent vows that the experience had burned into him.

Yet even amidst the grim machinery of conflict, life found ways to reassert itself.

In 1917, Creutzfeldt married Claire Sombart, the daughter of renowned sociologist and economist Werner Sombart. Claire brought a kind of steady warmth into his battered life, a reminder that beyond the noise and ruin, there remained the possibility of building something good, something lasting.

Together, they dreamed of a future beyond the war—a future filled with family, learning, and healing. It was a fragile dream, but it was enough to sustain them.

As the conflict dragged toward its bitter end, Creutzfeldt's duties grew heavier. In 1918, he was dispatched to the crumbling Ottoman Empire as part of the Military Mission staff. There, among the war-torn landscapes of Turkey and the Balkans, he organised hospital transports, tended to refugees, and served as a liaison officer to the Entente in Constantinople.

The work was endless, exhausting, and often heartbreaking. Refugee camps swelled with the displaced and the dying; disease festered unchecked. Still, Creutzfeldt's resilience never ceased. For his service, he was awarded the Iron Cross First and Second Class, and other military honours—acknowledgements of duty carried out under almost impossible conditions.

When the armistice was finally signed, bringing the bloodshed to a grudging halt, Creutzfeldt was interned in Turkey, —a final indignity that seemed to encapsulate the war's bitter ironies.

It was not until June 1919 that he returned to Germany. He was 34 years old, no longer the bright, eager student who had once peered down a microscope with untroubled eyes. Five years of loss, of conflict, of witnessing death on a scale previously unimaginable had carved deep furrows into his soul.

And yet, for all he had endured, Creutzfeldt refused to let those years define him. The quiet determination that had carried him through battlefields and icy waters now drove him to rebuild the life he had put on hold.

Returning to research felt like coming home. With his career back on track, Creutzfeldt turned his attention to the most pressing questions in psychiatric and neurological science, determined to make up for the years lost to war.

Among the preserved papers and specimens awaiting him was Bertha Elschker case. Her brain had been carefully sectioned and stored before he was called away to war, and now, with steadier hands and a changed heart, he examined it anew.

What he saw astonished him.

Bertha's brain revealed a series of severe and complex findings. Entire regions showed a disintegration not only of neurons, but of the surrounding support structures. Glial cells, normally quiet custodians of the brain, had surged into action, forming clusters and webs in an attempt to repair the damage.

The loss was most striking in the frontal lobe—the seat of personality, reasoning, memory, and emotional nuance. It explained the wild mood swings, the speech disturbances, the vacant gaze that had once unsettled even Alzheimer. But the devastation did not stop there. The basal ganglia, the thalamus, the cerebellum—each held its own scars, painting a picture of a disease that moved through the brain like a slow fire, sparing nothing.

Under magnification, the boundaries between health and decay blurred disturbingly. The transition was not abrupt, but gradual, as though the illness had crept forward cell by cell, fibre by fibre. Some neurons bulged grotesquely, their shapes distorted almost beyond recognition. Others had shrunk, their outlines collapsing inwards. Where cells had died, spidery formations bloomed in their place—a sign of the brain's futile attempts at repair. Even the blood vessels appeared altered, as though the very infrastructure of life within her skull had been sabotaged from within.

Among the altered tissue, Creutzfeldt spotted something curious—inclusion bodies, faint and minuscule, nestled within the degenerating cells. These might have been overlooked by a less attentive eye, but for Creutzfeldt, they held meaning. They hinted at the origins of the disease, perhaps even a clue to its cause.

By the time he completed his study, he had come to a quiet conclusion: this was not simply a case of mental illness. It was a neurological catastrophe. One that played out not just in symptoms, but in structure, in architecture, in the very fabric of the brain itself.

His findings were published in 1920. The report caused a stir. It described, for the first time, a recognisable pattern of degeneration linking clinical symptoms to pathological change. This was not speculation, not theory;— it was evidence. Creutzfeldt had documented a new category of brain disease, one that transcended psychiatry and stood at the threshold of something larger.

The publication marked more than a scientific breakthrough. It was his re-entry into a field that had changed while he was away, and his signal to the world that he had returned—not simply as a survivor, but as a thinker, a physician, and a mind to be reckoned with.

Among those who took notice of Creutzfeldt's publication was Alfons Jakob—a distinguished neuropathologist in his own right, and, like Creutzfeldt, a protégé of Alois Alzheimer. While their paths had diverged—Jakob remaining in Munich, Creutzfeldt having passed through Kiel, Breslau, and beyond—their intellectual roots ran deep into the same soil.

Jakob had risen to direct the neuropathology laboratory at Hamburg University, and it was there, sifting through cases that blurred the boundaries of neurology and psychiatry, that he began to identify more patients who displayed the same devastating symptoms. Progressive mental and physical decline, profound personality changes, and a rapid loss of motor control. These individuals would start by withdrawing from social interactions and becoming anxious or irritable. Over time, their speech slowed, their movements became unsteady, and they eventually became bedridden, succumbing to infections or other complications.

Post-mortem, Jakob saw the same signs Creutzfeldt had described: regions of brain tissue ravaged by degeneration, neurons replaced by glial scarring, and a distinctive sponge-like texture in key areas—a pocked and perforated architecture where structure had once been firm.

These were not isolated anomalies. Over the following years, Jakob identified multiple patients who mirrored the profile of Creutzfeldt's Bertha. The details varied—some declined over months, others over a few brief years—but the underlying pathology was unmistakable.

Jakob saw the link immediately. Though the two men had worked independently, their findings described the same devastating disease. Without collaborating directly, they had arrived at complementary truths— each man sketching a different angle of the same dark figure in the neurological landscape.

Together, their names would come to define one of the most haunting disorders of the 20th century: *Creutzfeldt-Jakob disease.*

Their findings marked a turning point, not just in the study of neurodegeneration, but in how medicine began to draw lines between the physical and the mental, the anatomical and the behavioural. In recognising the brain's capacity to shape—and destroy—the self, they helped lay the foundation for a more integrated vision of mind and body.

From 1924 to 1938, Creutzfeldt continued to build an illustrious career at the *Nervous Disorders Clinic* at the *Charité* in Berlin. Under the mentorship of renowned psychiatrist Karl Bonhoeffer, he thrived in an intellectually stimulating environment.

His research broadened in scope. He delved into histopathology and neurodegenerative disease, focusing on the microscopic anatomy of the brain and the vital, often overlooked role of microglial cells. These studies positioned him not just as a skilled observer but as a leader in the emerging field of neurological medicine. His work appeared in respected journals, quietly establishing his name among Europe's most thoughtful voices in neuroscience.

But Creutzfeldt was not only a researcher. He thrived as a teacher, pouring his energy into lectures on neuroanatomy and psychiatric pathology. He had a gift for explanation, and his classrooms— filled with the next generation of physicians,—became crucibles of modern neurological thinking. In 1927, he was promoted to Senior Assistant Physician, a reflection of both his academic authority and the deep respect he commanded among peers and students alike.

Yet, as his career soared, dark clouds gathered on the horizon. The rise of the National Socialist Party in Germany loomed, threatening to disrupt not only his work but the very principles of science and humanity that had guided him.

ⵣ

The rise of the National Socialist Party in 1933 brought a creeping shift—one that would soon upend every corner of German public life. Institutions, professions, and even private conversation began to bend under the weight of ideology. Science and medicine were swiftly conscripted into the regime's vision. Eugenics, racial theory, and militarism became not just political slogans, but research priorities.

Physicians were encouraged—and soon expected—to join party-aligned organisations. Participation signalled patriotism; refusal invited scrutiny. The German Doctors' Union, tightly linked with the Nazi state, grew to encompass tens of thousands of medical professionals. For many, membership was a career necessity. For others, it was a moral surrender.

Creutzfeldt resisted.

Unlike many of his colleagues, he declined full party membership, despite the pressure. His name appeared as an applicant—a bureaucratic formality, perhaps—but he never crossed the threshold into official affiliation. Privately, he found the regime's rhetoric grotesque and its methods abhorrent. He kept his distance, walking a precarious line between silent defiance and self-preservation.

Yet silence came at a cost. Navigating the new Germany required compromise, and Creutzfeldt, like so many, had to weigh ethics against survival. He became an associate member of the SS—not a militant or ideologue, but an affiliate who paid dues, made no public statements, and kept to his clinic. It was a paper shield, offering just enough protection to continue his work unimpeded.

Whether the decision was calculated, reluctant, or simply inevitable remains unclear. What mattered was that it bought him space—to think, to teach, to continue.

In 1938, he accepted a new post: Director of Psychiatry and Neurology at *Kiel University*. It was a position of prestige and responsibility, but also one that placed him ever closer to the heart of a regime he refused to embrace.

Creutzfeldt's quiet resistance did not go unnoticed. Those who knew him saw the distance he kept from the regime's machinery—a deliberate abstention from the loud, vulgar displays of loyalty that had become common currency in Nazi Germany. He avoided the rituals, the slogans, the easy comforts of conformity. Yet he understood all too well the dangers of open defiance. In a state where suspicion could be lethal, reckless protest would have endangered not only himself but his family, his colleagues, and the very patients he was trying to protect.

Instead, Creutzfeldt chose a path of careful navigation. He immersed himself in his work, presenting an outward professionalism that gave no cause for alarm, while inside, he balanced every action on the narrow edge between principle and survival.

As Director of the *Psychiatric and Neurological Clinic* at *Kiel University*, he oversaw a wide realm of responsibility—patient care, medical training, scientific research. After years of steady dedication, the role brought him stability, influence, and the space to continue his mission in medicine. Beyond his official duties, he remained a passionate teacher, mentoring young physicians with a seriousness that reflected his belief in knowledge as a form of resistance. He also maintained a private practice, a quiet but necessary source of income at a time when even professional standing offered little guarantee against hardship.

When World War II broke out in 1939, it unleashed immense challenges across Germany, and the *Psychiatric and Neurological Clinic* at Kiel was no exception. Resources dwindled; medical staff were conscripted; the demands of treating wounded soldiers never abated. Bombing raids devastated Kiel, battering the clinic's infrastructure and forcing Creutzfeldt and his remaining team to adapt under increasingly desperate conditions.

Yet through the hardship, Creutzfeldt endured. He worked ceaselessly to ensure his patients received care and his students continued to learn, refusing to let the chaos of war dismantle the clinic's reputation as a centre of psychiatric and neurological excellence.

But the war brought more than logistical trials. It carried deep personal and professional turmoil.

Creutzfeldt's close friend and former mentor, Karl Bonhoeffer, suffered a shattering loss when two of his sons and two sons-in-law were executed for their part in the anti-Nazi resistance. The weight of this tragedy bore down heavily. Creutzfeldt mourned privately with Bonhoeffer, yet the grief cracked something open within him. His lectures, once strictly academic, began to take on a sharper edge. He spoke out more plainly, condemning Hitler and his associates as murderers—his words cutting through the brittle, controlled silence that gripped Nazi Germany. It was an act of enormous risk at a time when even whispered dissent could cost lives.

Remarkably, Creutzfeldt's position remained intact, protected by an unlikely alliance. Former naval comrades from the First World War—now in high-ranking military posts—shielded him from reprisals. This "old boys" network, built on bonds forged through hardship decades earlier, provided a rare buffer. It allowed him to continue as director of the clinic, even as he rejected Nazi Party membership—one of only two professors at Kiel to do so, a decision that underscored the quiet steadfastness of his character.

Yet even this protection had limits. The regime's reach was vast, and by 1941, the horrors of the Nazi euthanasia programme, *Aktion T4*, had seeped into the very fabric of psychiatric medicine. Under *T4*, patients with psychiatric and neurological disorders were systematically targeted for extermination, their lives dismissed as "unworthy of life." Public protests forced the programme's official suspension that same year, but the killings persisted in secret, scattered through institutions that had learned how to mask atrocity beneath layers of bureaucracy.

For Creutzfeldt, *Aktion T4* was an abomination. It was a violation of everything his life in medicine had stood for: healing, understanding, and the irreducible value of human life.

It was against this grim backdrop that Helmut F. entered his world.

Diagnosed with schizophrenia and sentenced to confinement by a military court, Helmut's fate hung by a thread, caught within the ruthless machinery of the euthanasia regime. When he arrived under Creutzfeldt's care in 1941, his future was all but written.

For Creutzfeldt, Helmut was more than another patient. He was a test—a collision between medical duty and moral conviction. In a world where compliance had become the price of survival, Creutzfeldt would have to choose how far he was willing to go, and what dangers he would dare to invite, in order to save a single life.

Creutzfeldt first encountered Helmut at the clinic, where the young sailor's fate already seemed sealed by a damning schizophrenia diagnosis. But to Creutzfeldt's trained eye, the symptoms did not align with such a verdict. Instead, he recognised signs of a psychopathic personality—a crucial distinction at a time when a label of schizophrenia carried far graver consequences.

Acting in his role as consulting psychiatrist, Creutzfeldt quietly challenged the original assessment. His first step was measured: he approached Dr Elste, the Naval Senior Assistant Physician who had made the diagnosis, requesting a revision to the expert report. When his inquiry was met with silence, Creutzfeldt made an uncharacteristically bold move—bypassing the usual chain of command, he reported the case directly to the Navy Medical Office in June 1941.

The gamble worked. The court agreed to reopen Helmut's case and assigned Creutzfeldt to provide a new expert opinion.

In his reassessment, Creutzfeldt argued unequivocally that Helmut exhibited no signs of schizophrenia or any other serious mental illness. He diagnosed a psychopathic personality instead—a less stigmatising classification—and crucially emphasised that there were no grounds to declare Helmut mentally incapacitated under §51 of the Reich Penal Code. Without this protection, Helmut would almost certainly have been swept into the euthanasia programme.

Another expert later corroborated Creutzfeldt's findings, further strengthening Helmut's tenuous hold on life.

For Helmut, who had learned of the euthanasia programme during his earlier time at Zschadrass, Creutzfeldt's intervention meant salvation. He understood all too well what was at stake. To be marked with psychiatric illness was to live on the brink of extermination.

When Creutzfeldt informed him of the new findings, the young sailor's relief was visible. Helmut thanked him earnestly, speaking of how the doctor had "removed the burden of paragraph 51 from his shoulders." In a world indifferent to individual fates, this moment of human connection—fleeting, fragile, but genuine—mattered the most.

For Creutzfeldt, saving Helmut was not only an act of professional integrity but a quiet act of resistance. Yet he knew that every such defiance came at a cost. Even as he secured Helmut's survival, the dangers gathering around him grew heavier.

As Creutzfeldt negotiated these professional perils, the war's reach tightened around his personal life with devastating force. In 1942, his eldest son Harald made a decision few dared: abandoning his post in the navy, Harald disappeared into the underground resistance movement in the Netherlands. It was an act of extreme defiance, a betrayal in the eyes of the regime, and Creutzfeldt could only imagine the risks his son now faced, alone and hunted. The fear lingered constantly—unspoken, inescapable.

Then, in 1943, the war struck even closer. Claire, Creutzfeldt's wife and mother to their children, was arrested by the *Special Court of Kiel*, accused of making "malicious statements" against Nazi leaders. In a state where words could be fatal, her charges were grave: undermining public trust in the Reich's leadership.

Claire was sentenced to four months in prison—a term that might easily have been far harsher. The *Sondergericht*, notorious for dealing ruthlessly with political offences, had threatened to send her to a concentration camp. Only the intervention of Hans Burger-Prinz, a lawyer and old colleague from Hamburg, succeeded in reducing her sentence.

Even so, imprisonment was no ordinary ordeal. During an air raid on the prison, chaos erupted. Amidst the confusion, Claire seized her chance. She escaped and vanished into hiding for the remainder of the war.

Her absence left a hollow space within the family—a daily reminder of how precarious, how breakable, their existence had become under the regime's unyielding gaze.

For Creutzfeldt, the walls were closing in.

His family's ordeals—Harald's desertion into the resistance, Claire's arrest and imprisonment—exposed the brutal reality of life under a regime that spared no one. His network of colleagues and former naval comrades still offered some protection, but he knew instinctively that it could only stretch so far. Each decision he made, each word he spoke, felt like a tightrope walk over a chasm where a single misstep could bring ruin. For the first time, he fully grasped the precariousness of his existence. Survival demanded constant vigilance—a delicate balancing act between quiet defiance and outward compliance.

Meanwhile, the horrors of the Nazi euthanasia programme seeped deeper into the fabric of psychiatric practice. By 1944, the programme's reach extended across Germany, ensnaring even institutions like the clinics of *Kiel University*. Creutzfeldt's clinic was not tasked directly with reporting patients, but its role in diagnosing and treating acute psychoses placed it within the network of facilities complicit in-patient transfers. Transfers, once routine, now often meant a one-way journey for patients deemed "unfit"—those living with epilepsy, severe neurological disorders, or chronic psychiatric illnesses.

The scale of the killings became impossible to ignore. In September 1944, 698 patients from the overcrowded *Schleswig Hospital* were transferred to the *Provincial Asylum* at Obrawalde—among them, 53 children and adolescents, victims of what became known as "wild euthanasia." Few, if any, survived. Once delivered to places like Obrawalde, patients faced near-certain death, often by starvation, lethal injection, or other clinical methods masked as care.

Creutzfeldt watched these events with growing unease. He understood the fate that awaited many of those who left his care, and though he could not openly resist the regime, he worked quietly to shield as many lives as he could. Within the clinic, he delayed transfers, retained vulnerable patients longer than officially necessary, and adjusted diagnoses to less stigmatised categories.

These quiet acts of resistance were not invisible. Between 1938 and 1944, the percentage of patients at Kiel with conditions targeted by euthanasia criteria dropped sharply—from over 20% to less than 9%. The reduction was no accident. It was the result of Creutzfeldt's deliberate efforts to protect those trapped within a system designed to erase them.

Yet he knew the limits of what he could do. Claire's imprisonment, and the execution of Karl Bonhoeffer's sons, remained constant reminders that even the most cautious dissent could invite lethal reprisal. These grim lessons reinforced his strategy: to act where he could, but never in a way that would draw the spotlight of authority.

Years later, in September 1945, reflecting on the war's aftermath, Creutzfeldt would write, "I am a firm opponent of killing mentally ill patients."

It was a simple statement, free of grandeur—yet behind those few words lay the quiet defiance that had guided him through one of history's darkest chapters.

⚕

After World War II, Hans Gerhard Creutzfeldt emerged as a central figure in Germany's academic restoration. In 1945, he was appointed the first post-war *Rector of Kiel University*—a reflection of the respect he commanded for his political neutrality and unbroken integrity during the Nazi era.

The university he inherited was a shattered shell of its former self, ravaged by years of war and ideological corrosion. The task of rebuilding it fell squarely on Creutzfeldt's shoulders. He threw himself into the work, negotiating tirelessly with local officials and British occupation authorities to secure resources and re-establish the university on its original grounds. Under his leadership, the slow resurrection of Kiel's academic life began. But it was far from a smooth road.

Disputes quickly flared over the British-led denazification process, especially concerning the admission of former military officers into academic programmes. Creutzfeldt advocated for a balanced approach—one that sought justice without abandoning pragmatism—but tensions mounted. By 1946, the controversy reached a breaking point. He was dismissed by the British military government, a bitter blow after all he had worked to rebuild.

Creutzfeldt accepted the decision without public protest. Rather than retreat into resentment, he returned quietly to his old work, turning his full attention once more to medicine.

Resuming his position at the *Psychiatric and Neurological Clinic* in Kiel, he set about modernising the institution. New techniques— electroencephalography, neuroradiology—were introduced despite post-war scarcity, ensuring the clinic remained at the forefront of neurological care. Until his retirement in 1953, Creutzfeldt remained a driving force in German medicine, leaving behind a revitalised clinic and a legacy of resilience in one of the country's most turbulent eras.

After retiring, he relocated to Munich, intending to devote his remaining years to neuropathological research. Age and declining health slowed him, but not entirely. He continued to revisit his earlier work, contributed to smaller studies, and found quiet satisfaction in mentoring younger generations. He remained engaged with the science that had shaped his life.

However, even in retirement, shadows lingered. One final controversy drew him back into public life: the Heyde/Sawade affair.

Werner Heyde, a former SS officer and one of the chief architects of the Nazi euthanasia programme, had evaded justice, assuming the false identity of "Dr Fritz Sawade." In 1954, during a forensic dispute, Creutzfeldt noticed irregularities in Sawade's credentials. Quietly, methodically, he pieced together the truth.

It was a chilling discovery: a man responsible for orchestrating mass murder was living freely, practising medicine under a stolen name.

Determined to act, Creutzfeldt proceeded with characteristic caution. He reported Heyde's identity to the authorities, setting in motion investigations that would eventually expose one of post-war Germany's most notorious fugitives. While his actions were pivotal, he avoided the glare of public accusation, choosing a path shaped by the same quiet moral firmness that had guided him during the war.

The scandal unfolded long after his active career had ended, but it served as a stark reminder of the ethical battles that had marked his life's work.

Despite controversy and failing health, Creutzfeldt's contributions to medicine continued to be honoured. In 1955, he was named an Honorary Senator of Kiel University—a tribute to a lifetime of achievement and to the role he had played in rebuilding the institution from ruins.

On a cold December day in 1964, Hans Gerhard Creutzfeldt passed away in Munich at the age of 79. His death marked the end of a remarkable life, but his story did not end there.

In the decades that followed, historians and researchers would scrutinise his actions during the Nazi era, debating the extent of his resistance and the limits of what he could have done.

Could he have done more to protect patients caught in the euthanasia program? Were his subtle acts of defiance enough, given the position and influence he held? These questions linger, unanswered.

Perhaps they are the wrong questions to ask. Creutzfeldt's story forces us to grapple with a more challenging, more personal question: what would we have risked in his place? Could we have balanced moral conviction with survival during one of history's darkest periods, where every action carried the weight of life and death?

What is Creutzfeldt-Jakob disease?

Creutzfeldt-Jakob disease (CJD) is a rare brain disorder that disrupts brain function. It belongs to a group of illnesses known as prion diseases, named after the abnormal proteins—called prions—that cause the condition. These prions trigger healthy proteins in the brain to misfold and malfunction, leading to widespread brain damage.

As the disease progresses, brain cells die, and tiny sponge-like holes form in the brain tissue, giving it a distinctive appearance under the microscope. This damage happens rapidly, and the symptoms of CJD can worsen within just a few months. Early signs often include memory problems, difficulty walking or moving, and changes in personality, such as irritability or withdrawal. As the disease advances, confusion and severe mental decline set in, along with the loss of coordination and control over the body.

In its later stages, CJD is catastrophic. Individuals lose the ability to communicate, care for themselves, or respond to their surroundings. Sadly, the disease is always fatal, and there is currently no cure. While CJD is extremely rare, its impact on those affected—and their loved ones—is devastating.

CHAPTER 19

Myths, Madness, Ticks and Tics

AS WE REACH the final chapter of our journey through the stories behind medical eponyms, let me invite you to explore a fascinating alternative: names inspired by mythology, animals, geography, and even modern-day social media. While researching this book, I stumbled upon many such examples, each deserving its moment in the spotlight. These terms remind us that the language of medicine is rich, imaginative, and alive with history. Together, let's take a closer look at some of the most intriguing.

Greek Mythology & Medicine

Greek mythology is a vast and ancient treasury of powerful stories, passed down through centuries, shaping our language, imaginations, and ways of seeing the world. Words like *chaos*, *echo*, *fury*, and *panic* owe their origins to these myths, woven so deeply into our everyday speech that we barely notice them. In medicine, too, this tradition endures. The names of anatomical structures still echo the legends of gods, heroes, and monsters—binding ancient storytelling to modern science.

Take, for instance, the Achilles tendon. Sometimes called the heel cord, it is the thickest and strongest tendon in the human body, anchoring the powerful calf muscles—the gastrocnemius, soleus, and plantaris—to the heel bone. It allows us to walk, run, jump, and stand tall. Yet its name reminds us of fatal weakness.

Achilles, the great Greek warrior of Homer's *Iliad*, was the son of the mortal king Peleus and the immortal nymph Thetis. Desperate to protect her child, Thetis dipped Achilles into the river Styx, whose waters were said to grant invincibility. But as she held him by the heel, that small spot remained untouched—and vulnerable. Achilles grew into an unmatched fighter, seemingly impervious to harm, until the day Paris, a Trojan prince, struck his heel with an arrow, ending his life. Thus, the term "Achilles' heel" entered language as a metaphor for a single, devastating weakness amid otherwise great strength.

Then there is the atlas—the first cervical vertebra, resting just beneath the skull. Though small, it bears an immense load, supporting the head and enabling nodding and rotation. Its name honours Atlas, the mighty Titan who, after the Titans' war against the Olympian gods, was condemned by Zeus to hold up the heavens for eternity.

Atlas's punishment became one of mythology's most enduring images: a figure straining beneath the immense weight of the skies. Later, as the Renaissance imagined the Earth as a globe, Atlas was reimagined carrying the world itself—an image that now perfectly aligns with the role of the atlas bone. This small bone quite literally supports the skull—a world of another sort—quietly and ceaselessly.

And then there is Medusa—a name that conjures both terror and pity. Once a mortal woman renowned for her beauty, Medusa's life took a tragic turn when the goddess Athena cursed her. Medusa's lustrous hair was transformed into a tangle of writhing snakes, and her gaze became so dreadful that it turned onlookers to stone. In myth, Perseus, aided by divine gifts, beheaded Medusa; yet even severed, her head retained its petrifying power. Athena mounted it on her shield, using it both as a weapon and as a symbol of protection.

In medicine, Medusa's legacy lives on in the term *caput medusae*. This describes a striking clinical sign of portal hypertension (increased pressure in the veins that carry blood from the intestines to the liver). This condition, often seen in patients with liver cirrhosis, causes the veins around the umbilicus to distend and radiate outward, resembling Medusa's snake-like hair. For clinicians, the imagery is hauntingly apt, symbolising the intersection of myth and medicine.

Greek mythology doesn't just lend its voice to anatomical structures. It also offers powerful metaphors for human behaviour.

Consider the story of Narcissus, a youth of extraordinary beauty who fell hopelessly in love with his own reflection in a pool of water. Unable to tear himself away, he wasted away by the water's edge, dying of longing for himself. From this tale, we derive the concept of *narcissism*: an excessive preoccupation with oneself, often at the expense of meaningful relationships and reality.

In modern psychoanalysis, narcissism evolved into a framework for understanding personality traits marked by self-absorption, fragile self-esteem, and the desperate need for validation. Narcissus's story serves as a timeless reminder of the dangers of unchecked vanity and self-obsession.

Another powerful myth that has left its mark on psychology is the tragedy of Oedipus. Born to King Laius and Queen Jocasta of Thebes, Oedipus was cursed from infancy: a prophecy foretold he would kill his father and marry his mother. To thwart fate, Laius abandoned his newborn son to die on a mountainside.

Yet destiny was not so easily defied. Rescued and raised by strangers, Oedipus grew up unaware of his true lineage. As a young man, he quarrelled with and killed a stranger on the road—his own father—and later married the widowed Queen of Thebes, Jocasta—his mother. When the horrifying truth emerged, Jocasta took her life, and Oedipus, overwhelmed by guilt and horror, blinded himself.

From this myth, Sigmund Freud developed the concept of the *Oedipus complex*, which describes a child's subconscious desire for the parent of the opposite sex and feelings of rivalry toward the same-sex parent. While Freud's theories have been reinterpreted over time, the *Oedipus complex* remains a potent example of how mythology can illuminate universal human experiences—identity, desire, and conflict—helping us make sense of the complexities of our inner worlds.

These stories—of Achilles, Atlas, Medusa, Narcissus, and Oedipus—remind us of the power of mythology. It is a deep reservoir of human meaning: of strength and frailty, of self-awareness and delusion, of the mind's capacity for both triumph and tragedy. By naming our bones, diseases, and psychological conditions after these ancient tales, we preserve their legacy and weave them into the very fabric of science and medicine.

Mad Cow Disease

It began with a farmer in the quiet English countryside in 1986, watching over his herd as he always did. The mornings were his favourite—dew glistening on the grass and his cows' soft lowing as they wandered lazily toward their feed.

But something wasn't right.

One of his prized cows, a reliable old girl he knew by name, seemed different. She stumbled slightly as she approached the trough, her gait unsure. The farmer frowned but dismissed it. Perhaps she was just tired, or maybe it was nothing at all.

Days passed, and the cow's behaviour became more troubling. Her gentle demeanour gave way to restless agitation. She startled at the slightest noise, kicking and thrashing as if warding off invisible threats. Her wide, panicked eyes held a look he couldn't quite describe—confusion, perhaps, or fear. Soon, she struggled to stand, her legs buckling beneath her.

The farmer had nursed sick animals before but this was something different. Something chaotic. The gentle patterns of farm life were unravelling before his eyes.

At first, it was an isolated tragedy. Then it spread.

Across the countryside, farmers watched helplessly as their herds began to exhibit the same terrifying symptoms. Staggering steps, trembling limbs, frantic outbursts—it was as if their cattle were losing their minds.

By the early 1990s, the mysterious illness had swept through herds across the United Kingdom, devastating farms and livelihoods. The name it earned, *"Mad Cow Disease,"* was both stark and evocative, capturing the chaos of the symptoms. Officially, it was called *Bovine Spongiform Encephalopathy* (BSE), but the sensational name stuck—a reflection of the madness it seemed to invoke in affected cattle.

The origins of this nightmare lay not in the fields but in the feed. To maximise efficiency, cattle were fed a meat and bone meal supplement—a mixture made from the ground-up remains of other animals, including sheep. It was a common practice at the time, one that seemed harmless. But it turned out to be disastrously flawed. Some of those sheep had been infected with *scrapie*, a prion disease.

Prions are biological anomalies—misfolded proteins that are neither alive nor dead yet wreak havoc on the brain. Resistant to heat, disinfectants, and most decontamination methods, they are the ultimate biological saboteurs.

When cattle ingested contaminated feed, these prions entered their bodies silently, triggering a destructive chain reaction. Normal proteins in the brain, coming into contact with their corrupted counterparts, twisted and deformed in turn. Over time, the brain became riddled with sponge-like holes, its structure collapsing from within.

The disease could lie dormant for years, hidden and silent, before the first signs emerged. But when it surfaced, the outcome was inevitable: death.

By 1993, the epidemic had reached its dreadful peak. More than 100,000 confirmed cases of BSE were recorded across Britain. Farmers mourned not just the economic loss but the animals they had raised and cared for like family. Entire livelihoods crumbled. Governments scrambled to contain the outbreak, but for many, it was already too late.

At first, the crisis was seen as a tragedy confined to the countryside—a catastrophe for farming communities, but one that seemed remote from urban life.

Then, in 1996, everything changed.

Health authorities announced that BSE had crossed the species barrier. Humans who consumed infected beef products were at risk of developing a new and fatal illness: *variant Creutzfeldt-Jakob Disease* (vCJD).

The announcement sent shockwaves through the public. Families who had enjoyed beef dinners for generations began to question every meal. Grocery store aisles emptied of beef products, and restaurants hastily rewrote their menus. The disease had leapt across the species barrier for the first time, turning a farming tragedy into a personal fear.

The link between BSE and vCJD wasn't just a scientific revelation but a bombshell. The thought of an invisible killer lurking in something as ordinary as a meal turned every plate of beef into a source of unease. Unlike the cattle, humans didn't show symptoms immediately. vCJD could incubate silently for years, working its way through the brain before the first signs emerged. And when they did, they were devastating.

In humans, the disease often begins with subtle changes— depression, anxiety, or mood swings. But these early signs soon led to trembling, unsteady movements, and a rapid decline in mental abilities. Victims, many of them shockingly young, often in their late 20s or early 30s, would lose the ability to speak or move as the disease progressed. Death usually followed within months of the first symptoms.

The public reaction was swift and overwhelming. Governments worldwide banned the import of British beef, and trust in food safety systems plummeted. In the UK, the agricultural sector was battered by a crisis unlike anything it had faced before. Entire herds were culled—not just the sick animals, but countless healthy ones too—as part of a desperate attempt to stop the disease from spreading.

For farmers, this wasn't just a financial loss—it was like losing members of their own family. These were animals they had raised from calves and fed, named, and cared for.

But the culling wasn't just driven by emotion but a scientific necessity. To prevent further transmission, governments banned the use of meat and bone meal in livestock feed. During slaughter, specific parts of the cow—such as the brain and spinal cord, where prions were most heavily concentrated—were carefully removed and destroyed. New surveillance programmes were rolled out, focusing especially on older cattle, where the disease was most likely to emerge.

Slowly, these efforts began to work. By the early 2000s, new cases of BSE in cattle had dropped dramatically, thanks to stricter regulations and better monitoring systems.

Still, the scars of the epidemic remained. The *Mad Cow* crisis served as a brutal lesson about the delicate balance between human ambition and nature's inscrutable laws. By forcing herbivores into a carnivorous diet in the name of efficiency, humanity had breached a natural boundary—and the cost was staggering. Not just in lives lost, but in the trust that once bound people to the food they put on their tables.

Lyme disease

In the autumn of 1975, the woods around Old Lyme, Connecticut, were ablaze with autumn's golden hues. Towering oaks and maples cast long shadows over a town that seemed idyllic in its quiet beauty. But beneath the picturesque landscape, something was majorly wrong.

Children who should have been running through the crisp air were instead confined indoors, their joints swollen and stiff. Adults, too, found their movements slowed by pain, their limbs gripped by what felt like arthritis. For a disease so often associated with old age, this sudden epidemic among the young and healthy was as baffling as it was terrifying.

Two mothers, each with children suffering from the mysterious illness, decided they couldn't wait for answers any longer. Frustrated by the lack of explanations, they contacted the *Connecticut State Department of Health* and *Yale School of Medicine*. Their calls for help did not go unanswered. Among the scientists who responded were Dr Allan Steere, a young rheumatology fellow, and his mentor, Dr Stephen Malawista. Together, they began what would become the first chapter in the story of *Lyme disease*, a name that would one day become synonymous with one of the most perplexing illnesses.

Dr Steere and his team launched a rigorous investigation into what was then called *"Lyme arthritis."* They travelled from home to home in Old Lyme and the surrounding towns, collecting patient histories and conducting examinations. The more they learned, the stranger the patterns became. Most affected families lived in wooded, rural areas rather than lively town centres. Entire families were falling ill, with clusters of cases appearing on single roads. A curious seasonal pattern also emerged: cases surged during the summer and early autumn, only to vanish with the arrival of frost.

But one clue stood out above all the rest.

Many patients recalled seeing a peculiar rash before their symptoms began—a red, expanding ring with a clear centre, eerily resembling a bull's eye. For some, the rash was accompanied by flu-like symptoms: fatigue, headaches, fever, and muscle pain. Dr Steere recognised the lesion from European medical literature, where it was called *erythema migrans*. In Europe, this rash had been linked to tick bites—though never to arthritis.

Could ticks be the missing piece of the puzzle?

As the investigation deepened, the disease revealed new layers of complexity. While swollen joints remained the most common complaint, some patients suffered far more alarming complications. A few developed facial paralysis, severe headaches, or confusion as the illness attacked their nervous systems. Others experienced irregular heart rhythms that threatened their lives.

This was no simple arthritis—it was a multi-systemic disease capable of wreaking havoc across the body in unpredictable ways.

The *Yale* team's suspicions about ticks grew stronger as they expanded their investigation into neighbouring regions. They discovered the illness was far more common east of the Connecticut River—an area dense with deer and other wildlife. The connection became undeniable: deer carried ticks, and ticks were spreading disease.

Yet it wasn't until 1982 that the final piece of the puzzle fell into place. Willy Burgdorfer, a Swiss-American microbiologist, identified the true culprit: a spirochete bacterium that would later bear his name, *Borrelia burgdorferi*. This corkscrew-shaped organism thrived inside the gut of *Ixodes scapularis*—the black-legged deer tick.

The tick's life cycle quickly became a critical focus of study. Over the course of two years, the tick progresses through three stages: larva, nymph, and adult. In its larval stage, it feeds on small animals like mice, which serve as reservoirs for *Borrelia burgdorferi*. As larvae mature into nymphs, they often bite humans during the warmer months—spring and summer—when outdoor activities are at their height. No larger than a poppy seed, nymphs are so tiny that their bites often go completely unnoticed. Adults, which are larger and more easily spotted, typically feed on deer, where they mate and complete their lifecycle. This intricate interplay between ticks, mice, deer, and humans provided the perfect pathway for spreading the bacterium.

The stealthiness of the tick explained why so many patients failed to recall being bitten. The bacterium's own ability to evade the immune system added another layer of mystery. Once inside the body, *Borrelia burgdorferi* could burrow deep into joints, the nervous system, and even the heart, triggering widespread damage.

With the discovery of the bacterium, the illness was officially named *Lyme disease*, forever tying it to the small Connecticut town where it had first been described.

The identification of *Borrelia burgdorferi* marked a turning point, but the battle was far from over. *Lyme disease* proved to be a master of disguise. Symptoms varied wildly from patient to patient, and not everyone developed the telltale bull's-eye rash. For many, the early signs—fatigue, joint pain, flu-like symptoms—were vague, easily overlooked or misdiagnosed. Left untreated, the disease could progress to its late stages, causing debilitating arthritis, chronic neurological problems, and severe heart inflammation.

Fortunately, treatment brought hope.

Early in their research, *Yale* scientists had tested antibiotics on patients and found them to be effective, particularly when administered soon after infection. A pivotal 1980 study confirmed that antibiotics not only cleared the rash more quickly but also prevented the development of arthritis. Today, antibiotics remain the cornerstone of *Lyme disease* treatment. Patients treated in the early stages typically recover fully, though those who reach the later phases of the disease often require longer and more intensive therapy.

Still, challenges remain.

Diagnosing *Lyme disease* is not always straightforward. Many patients never notice the tick bite or the rash, and blood tests for *Lyme antibodies* can be inconclusive, especially in the early stages of infection. Public health campaigns now emphasise prevention: wearing protective clothing in wooded or grassy areas, using insect repellents, and conducting thorough tick checks after outdoor activities. Simple measures like these have become essential tools in controlling the disease's spread.

TikTok Tics

In 2020, the world stood still, paralysed by the isolation and uncertainty of the COVID-19 pandemic. Fear and anxiety seeped into every aspect of daily life, exacerbated by lockdowns that confined millions to their homes. Social interactions were replaced with digital connections, and *TikTok's* short-form video platform emerged as a lifeline for entertainment, creativity, and a sense of community.

But as the app's popularity surged, an unexpected phenomenon unfolded.

Sophia, a bright and imaginative 14-year-old, was one of the millions who found solace in *TikTok* during the pandemic. Living in a quiet town with little to do and few friends nearby, she spent countless hours scrolling through videos, captivated by quirky dances, lip-syncs, and comedic skits. For Sophia, *TikTok* became her escape—a window to a world beyond the four walls of her bedroom.

But as the days turned into weeks, she began noticing something unsettling—a change in her body she couldn't explain.

It started with small, jerky movements in her arms and legs. At first, she brushed them off, thinking they were just nerves or the result of spending too much time sitting at her desk. But within weeks, the movements escalated. Her arms would suddenly punch the air, her legs would jerk uncontrollably, and she found herself shouting phrases she couldn't control in a voice that didn't feel like her own. Even the slightest triggers—a flashing light, the sound of a whistle—could set off these episodes. At times, Sophia would collapse into fits of trembling, her hands shaking violently as if they were no longer her own.

Her family watched helplessly, their worry mounting with each passing day. Sophia's doctors were equally perplexed. Her symptoms resembled *Tourette syndrome*, a neurological disorder characterised by involuntary tics. But there was something different about Sophia's case—something that didn't quite fit the familiar patterns.

As her family searched desperately for answers, scouring the internet and consulting specialist after specialist, they realised Sophia was not alone.

Across the globe, similar reports began to surface. In clinics and hospitals, neurologists and psychiatrists observed a startling rise in young girls—most of them teenagers— suddenly developing severe, complex tics. These were not just scattered incidents but part of a global wave, a phenomenon that defied easy explanation.

As physicians scrambled to understand what was happening, a connection began to take shape.

Many of the affected teens, like Sophia, had spent significant time on *TikTok*. Even more strikingly, they were avid followers of influencers who openly shared their lives with *Tourette syndrome*. These influencers, some with millions of followers, frequently posted videos demonstrating their tics—repetitive movements, vocal outbursts, even coprolalia, the involuntary utterance of obscene words.

Sophia admitted that she, too, followed a popular British *TikTok* creator whose tic videos had gone viral. At first, she mimicked the movements and phrases she saw online.

But somewhere along the way, what had started as harmless imitation slipped beyond her control. The tics became real—unbidden, unstoppable—until *TikTok*, once her comfort, became a strange and unsettling mirror, reflecting back symptoms she could no longer suppress.

Clinicians were initially confounded by the similarities to *Tourette syndrome* but soon realised there were crucial differences. *Tourette's* typically begins in early childhood, with simple motor tics that gradually evolve, often affecting the face, eyes, and head.

By contrast, *TikTok* tics appeared suddenly, almost overnight, in adolescents who had no prior history of tic disorders. The tics were more severe, more complex, often involving dramatic motor movements like punching or kicking—and strikingly specific vocalisations.

In Sophia's case, her outbursts included phrases and gestures identical to those demonstrated by her favourite *TikTok* influencer. Unlike classical *Tourette's*, *TikTok* tics showed no consistent progression over time and often erupted in bursts triggered by environmental stimuli like flashing lights or sudden sounds.

The phenomenon fit the description of *functional neurological symptom disorder*, a condition where psychological stress manifests as physical symptoms.

Sophia's pre-existing struggles with depression and anxiety made her particularly vulnerable. The pandemic's isolation, fear, and uncertainty amplified these emotions, creating fertile ground for a phenomenon driven by social influence.

TikTok became the conduit, spreading tic-like behaviours across borders and turning what might once have been a localised outbreak into a global trend.

The story of *TikTok tics* is not as unusual as it might seem. It is, in fact, the modern iteration of a phenomenon as old as humanity itself: *sociogenic illness*, where symptoms spread through groups without an identifiable physical or environmental cause. Far from being "faked," these symptoms are genuine and distressing and arise from psychological factors like fear, stress, and suggestibility.

Sociogenic illnesses do not occur in isolation—they thrive in specific cultural and social conditions. The COVID-19 pandemic, with its unprecedented levels of stress and anxiety coupled with the omnipresence of social media, created the perfect storm for just such an outbreak.

History offers striking parallels to this modern phenomenon.

In 1518, the city of Strasbourg was gripped by the *Dancing Plague*, where hundreds of people inexplicably began dancing uncontrollably in the streets. It started with a single woman, Frau Troffea, whose relentless movements drew an anxious crowd. Soon, others joined her, dancing for days on end until some collapsed from exhaustion or even died.

Scholars today believe the outbreak was a collective psychological response to the stress of famine, disease, and religious oppression—a desperate and unconscious attempt to release pent-up anxiety.

Centuries later, in 1994, the *"Toxic Lady"* incident unfolded in a Californian hospital. Gloria Ramirez, a cancer patient, was rushed to the emergency room, where staff noticed a strange chemical odour emanating from her body. Within minutes, several healthcare workers began experiencing nausea, dizziness, and even temporary paralysis. Panic spread through the hospital, as theories of chemical exposure gripped the staff. Yet no definitive toxic source was ever found. Investigators ultimately concluded that fear and suggestion had triggered a sociogenic illness among the hospital team.

In 1995, following the deadly sarin gas attacks in Tokyo, orchestrated by the Aum Shinrikyo cult, another wave of sociogenic illness emerged. While the nerve agent killed 13 people and injured hundreds, thousands more—many of whom had no direct exposure—flooded hospitals, reporting symptoms like shortness of breath, dizziness, and nausea. The psychological trauma of the attack, coupled with the fear of chemical warfare, fuelled a massive collective response, overwhelming emergency services.

These historical cases reveal the extraordinary power of collective belief and psychological stress to shape human experience.

What makes *TikTok* tics unique is their digital nature. Social media, with its unparalleled reach and immediacy, has become a new catalyst for spreading sociogenic illness. Just as past sociogenic illnesses reflected their times—be it famine, toxic scares, or terrorism—modern platforms like *TikTok* have given rise to their own medical narratives, even earning a medical eponym in the process.

Acknowledgements

FIRST AND FOREMOST, I owe the biggest thank you to my sister, Grace. She had the patience (and tolerance!) to read through endless drafts, gently pointing out where I slipped back into medical jargon. Honestly, her ability to put up with me during this process deserves its own award.

To my partner, Jodie — thank you for being there every step of the way. Your support kept me going when the writing felt never-ending, and knowing you were in my corner made all the difference.

I'm also grateful to the team at Greenhill Publishing. They've backed me in every project we've worked on together, and their guidance has made the publishing world feel a lot less intimidating.

And of course, this book exists because of the brilliant, trail-blazing figures whose names have become woven into medical history. Their work continues to inspire and shape how we practice medicine today.

Finally, to you, the reader — thank you for joining me on this journey. The next time you come across a condition with an eponym, I hope you'll pause and be curious about the person behind it. Look beyond the white coat.

Works Cited

Introduction – What is a Medical Eponym?

Almagià, R. (2025). Amerigo Vespucci. In *Britannica*. Encyclopedia Britannica. https://www.britannica.com/biography/Amerigo-Vespucci

Ambrose, J. (1987, June 3). Caesar hailed— His salad's never tossed out. *Honolulu Star-Bulletin*.

Britannica, T. E. of E. (2024). Rudolf Diesel. In *Britannica*. Encyclopedia Britannica. https://www.britannica.com/biography/Rudolf-Diesel

Cappuzzo, B. (2008). Eponyms or descriptive equivalent terms? The question of scientific accuracy in medical discourse. *Esercizi, Miscellanea Del Dipartimento Di Scienze Filologiche e Linguistiche*, 2, 25–35.

Eponym, N. (2023). In *Oxford English Dictionary*. Oxford University Press. https://doi.org/10.1093/OED/1190617924

Hall, A. (2023, June 23). What is Tourette syndrome, the condition Lewis Capaldi has? *SBS News Australia*.

le Zotte, J. (2017, October 3). *When Cardigans Were Battle Attire*. Racked. https://www.racked.com/2017/10/3/16380180/cardigans-history

O'Dowd, N. (2023, June 19). *The English land agent who inspired the "boycott" in 19th-century Ireland*. Irish Central. https://www.irishcentral.com/roots/history/irish-invented-boycott

Okun, M. S., Mayberg, H. S., & DeLong, M. R. (2023). Muhammad Ali and Young-Onset Idiopathic Parkinson Disease—The Missing Evidence. *JAMA Neurology*, *80*(1), 5. https://doi.org/10.1001/jamaneurol.2022.3584

The Michael J. Fox Foundation. (n.d.). *Michael's Story*. The Michael J. Fox Foundation. Retrieved January 4, 2025, from https://www.michaeljfox.org/michaels-story

Zemeckis, R. (1985). *Back to the Future*. Universal Pictures.

Chapter 1 – The Rebellious Movement Disorder

Bloem, B. R., Okun, M. S., & Klein, C. (2021). Parkinson's disease. *The Lancet*, *397*(10291), 2284–2303. https://doi.org/10.1016/S0140-6736(21)00218-X

Gardner-Thorpe, C. (2010). James Parkinson (1755–1824). *Journal of Neurology*, *257*(3), 492–493. https://doi.org/10.1007/s00415-009-5440-8

Goetz, C. G. (1986). Charcot on Parkinson's disease. *Movement Disorders*, *1*(1), 27–32. https://doi.org/10.1002/mds.870010104

Kempster, P. A., Hurwitz, B., & Lees, A. J. (2007). A new look at James Parkinson's Essay on the Shaking Palsy. *Neurology*, *69*(5), 482–485. https://doi.org/10.1212/01.wnl.0000266639.50620.d1

Lewis, P. A. (2012). James Parkinson: The Man Behind the Shaking Palsy. *Journal of Parkinson's Disease, 2*(3), 181–187. https://doi.org/10.3233/JPD-2012-012108

Parent, A. (2018). A Tribute to James Parkinson. *Canadian Journal of Neurological Sciences / Journal Canadien Des Sciences Neurologiques, 45*(1), 83–89. https://doi.org/10.1017/cjn.2017.270

Parkinson, J. (2002). An Essay on the Shaking Palsy. *The Journal of Neuropsychiatry and Clinical Neurosciences, 14*(2), 223–236. https://doi.org/10.1176/jnp.14.2.223

Tyler, K. L., & Tyler, H. R. (1986). The secret life of James Parkinson (1755-1824). *Neurology, 36*(2), 222–222. https://doi.org/10.1212/WNL.36.2.222

Chapter 2 – Catch Your Breath

Grasha, K., & Tweh, B. (2016, May 26). At 96, Heimlich performs his own maneuver. *The Enquirer.*

Heimlich, H. J., & Patrick, E. A. (1990). The Heimlich maneuver. *Postgraduate Medicine, 87*(6), 38–53. https://doi.org/10.1080/00325481.1990.11716329

Jane Elliott. (2003, March 9). Heimlich: Still saving lives at 83. *BBC News.*

Mcfadden, R. D. (2016, December 17). Dr. Henry J. Heimlich, Famous for Antichoking Technique, Dies at 96. *International New York Times.*

Phillips, A. (2009, November 2). Dr. Henry Heimlich, Medical Innovator. *VOA News.*

Radel, C. (2016, December 17). Henry Heimlich, inventor of Heimlich maneuver, dies at 96. *The Cincinnati Enquirer.*

Roehr, B. (2017). Henry Heimlich. *BMJ*, j118. https://doi.org/10.1136/bmj.j118

WCPO Staff. (2016, December 18). Dr. Henry Heimlich, inventor of famed anti-choking Heimlich maneuver, dies. *WCPO 9 Cincinnati.*

Zengerle, J. (2007, April 24). The Bizarre Life and Times of the Inventor of the Heimlich Manuever. *The New Republic.*

Chapter 3 – Tics and Turmoil

Adanır, S. S., & Bahşi, İ. (2021). Life and works of Gilles de la Tourette (1857–1904). *Child's Nervous System, 37*(10), 2955–2958. https://doi.org/10.1007/s00381-019-04327-5

Krämer, H., & Daniels, C. (2004). Pioneers of movement disorders: Georges Gilles de la Tourette. *Journal of Neural Transmission, 111*(6), 691–701. https://doi.org/10.1007/s00702-004-0113-3

Lanska, D. J. (2018). *Jumping Frenchmen, Miryachit, and Latah: Culture-Specific Hyperstartle-Plus Syndromes* (pp. 122–131). https://doi.org/10.1159/000475700

Leckman, J. F. (2002). Tourette's syndrome. *The Lancet, 360*(9345), 1577–1586. https://doi.org/10.1016/S0140-6736(02)11526-1

Lees, A. J. (1986). Georges Gilles de la Tourette. The man and his times. *Revue Neurologique, 142*(11), 808–816.

Rickards, H., & Cavanna, A. E. (2009). Gilles de la Tourette: The man behind the syndrome. *Journal of Psychosomatic Research, 67*(6), 469–474. https://doi.org/10.1016/j.jpsychores.2009.07.019

Chapter 4 – Irish Charm

Jay, V. (1999). Dr Robert James Graves. *Archives of Pathology & Laboratory Medicine, 123*(4), 284–284. https://doi.org/10.5858/1999-123-0284-DRJG

Sawin, C. T. (1998). THEORIES OF CAUSATION OF GRAVES' DISEASE. *Endocrinology and Metabolism Clinics of North America, 27*(1), 63–72. https://doi.org/10.1016/S0889-8529(05)70298-X

Smith, T. J., & Hegedüs, L. (2016). Graves' Disease. *New England Journal of Medicine, 375*(16), 1552–1565. https://doi.org/10.1056/NEJMra1510030

Stokes, W. (1863). *Studies in Physiology and Medicine.* J. Churchill.

Taylor, S. (1986). Graves of Graves' disease, 1796-1853. *Journal of the Royal College of Physicians of London, 20*(4), 298–300.

Chapter 5 – A Mother to All

Anecdotesanaesthesia. (2019, November 2). *"The Phone booth Caper" Entertaining History about Virginia Apgar.* Forgotten Heroes of Anaesthesia. https://anecdotesanaesthesia. wordpress.com/2019/11/02/the-phone-booth-caper-entertaining-histo ry-about-virginia-apgar/

Apgar, V. (1958). EVALUATION OF THE NEWBORN INFANT-SECOND REPORT. *Journal of the American Medical Association, 168*(15), 1985. https://doi.org/10.1001/ jama.1958.03000150027007

Calmes, S. H. (1984). Virginia Apgar: a woman physician's career in a developing specialty. *Journal of the American Medical Women's Association (1972), 39*(6), 184–188.

Calmes, S. H. (2015). Dr. Virginia Apgar and the Apgar Score. *Anesthesia & Analgesia, 120*(5), 1060–1064. https://doi.org/10.1213/ANE.0000000000000659

Finster, M., Wood, M., & Raja, S. N. (2005). The Apgar Score Has Survived the Test of Time. *Anesthesiology, 102*(4), 855–857. https://doi.org/10.1097/00000542-200504000-00022

James, L. S. (1975). Fond memories of Virginia Apgar. *Pediatrics, 55*(1), 1–4.

Ray, A. R., Haines, D., & Grell, R. (2024). Virginia Apgar (1909-1974): The Mother of Neonatal Resuscitation. *Cureus.* https://doi.org/10.7759/cureus.61115

van Robays, J. (2015). The story of Virginia Apgar. *Obgyn 7,* 192–197.

Whonamedit? (n.d.). *Virginia Apgar.* Whonamedit? Retrieved December 10, 2024, from https://www.whonamedit.com/doctor.cfm/204.html

Chapter 6 – Smile

Duan, D., Goemans, N., Takeda, S., Mercuri, E., & Aartsma-Rus, A. (2021). Duchenne muscular dystrophy. *Nature Reviews Disease Primers, 7*(1), 13. https://doi.org/10.1038/ s41572-021-00248-3

Ekman, P., Davidson, R. J., & Friesen, W. v. (1990). The Duchenne smile: emotional expression and brain physiology. II. *Journal of Personality and Social Psychology, 58*(2), 342–353.

Hueston, J. T., & Cuthbertson, R. A. (1978). Duchenne de Boulogne and Facial Expression. *Annals of Plastic Surgery, 1*(4), 411–420. https://doi. org/10.1097/00000637-197807000-00009

MARANHÃO-FILHO, P., & VINCENT, M. (2019). Guillaume-Benjamin Duchenne: a miserable life dedicated to science. *Arquivos de Neuro-Psiquiatria, 77*(6), 442–444. https:// doi.org/10.1590/0004-282x20190044

Nelson, K. R., & Genain, C. (1989). Duchenne de Boulogne and the Muscle Biopsy. *Journal of Child Neurology, 4*(4), 315–315. https://doi.org/10.1177/088307388900400413

Parent, A. (2005a). Duchenne de Boulogne (1806–1875). *Parkinsonism & Related Disorders, 11*(7), 411–412. https://doi.org/10.1016/j.parkreldis.2005.04.004

Parent, A. (2005b). Duchenne De Boulogne: A Pioneer in Neurology and Medical Photography. *Canadian Journal of Neurological Sciences / Journal Canadien Des Sciences Neurologiques, 32*(3), 369–377. https://doi.org/10.1017/S0317167100004315

PEARCE, J. M. S. (1999). Some contributions of Duchenne de Boulogne (1806-75). *Journal of Neurology, Neurosurgery & Psychiatry, 67*(3), 322–322. https://doi.org/10.1136/jnnp.67.3.322

Images:

File:Guillaume Duchenne de Boulogne performing facial electrostimulus experiments. jpg. (2022, October 29). *Wikimedia Commons.* Retrieved January 6, 2025 from https://commons.wikimedia.org/w/index.php?title=File:Guillaume_Duchenne_de_Boulogne_performing_facial_electrostimulus_experiments.jpg&oldid=700261932.

File:Guillaume Duchenne de Boulogne performing facial electrostimulus experiments (3). jpg. (2017, September 15). *Wikimedia Commons.* Retrieved January 6, 2025 from https://commons.wikimedia.org/w/index.php?title=File:Guillaume_Duchenne_de_Boulogne_performing_facial_electrostimulus_experiments_(3).jpg&oldid=258720952.

File:Duchenne-FacialExpressions.jpg. (2023, November 22). *Wikimedia Commons.* Retrieved January 6, 2025 from https://commons.wikimedia.org/w/index.php?title=File:Duchenne-FacialExpressions.jpg&oldid=824177241.

Chapter 7 – A New Sound

Bloch, H. (1993). The inventor of the stethoscope: René Laennec. *The Journal of Family Practice, 37*(2), 191.

Davies, M. K., & Hollman, A. (1996). René Théophile-Hyacinthe Laennec (1781-1826). *Heart (British Cardiac Society), 76*(3), 196. https://doi.org/10.1136/hrt.76.3.196

Davies, M. K., & Hollman, A. (1997). Joseph Leopold Auenbrugger (1722-1809). *Heart, 78*(2), 102–102. https://doi.org/10.1136/hrt.78.2.102

Herzog, B. H. (1998). History of Tuberculosis. *Respiration, 65*(1), 5–15. https://doi.org/10.1159/000029220

Jay, V. (2000). The Legacy of Laënnec. *Archives of Pathology & Laboratory Medicine, 124*(10), 1420–1421. https://doi.org/10.5858/2000-124-1420-TLOLN

Kligfield, P. (1981). Laennec and the discovery of mediate auscultation. *The American Journal of Medicine, 70*(2), 275–278. https://doi.org/10.1016/0002-9343(81)90762-2

Roguin, A. (2006). Rene Theophile Hyacinthe Laennec (1781-1826): The Man Behind the Stethoscope. *Clinical Medicine & Research, 4*(3), 230–235. https://doi.org/10.3121/cmr.4.3.230

Sakula, A. (1981). R T H Laennec 1781--1826 his life and work: a bicentenary appreciation. *Thorax, 36*(2), 81–90. https://doi.org/10.1136/thx.36.2.81

Stone, M. J. (2005). Thomas Hodgkin: Medical Immortal and Uncompromising Idealist. *Baylor University Medical Center Proceedings, 18*(4), 368–375. https://doi.org/10.1080/08998280.2005.11928096

Yaqub, F. (2015). René Théophile Hyacinthe Laënnec. *The Lancet Respiratory Medicine, 3*(10), 755–756. https://doi.org/10.1016/S2213-2600(15)00374-4

Chapter 8 – A Stroke of Insight

Caplan, L. R. (2020). *C. Miller Fisher: stroke in the 20th Century.* Oxford University Press, USA. https://doi.org/10.1093/med/9780190603656.001.0001

Caplan, L. R., Mohr, J. P., & Ackerman, R. H. (2012). In Memoriam: Charles Miller Fisher, MD (1913-2012). *Archives of Neurology, 69*(9), 1208. https://doi.org/10.1001/archneurol.2012.1743

Fisher, C. M. (1945). Marlag 1941-44. *Canadian Medical Association Journal, 52*(3), 305–307.

Fisher, C. M. (2001). A Career in Cerebrovascular Disease: A Personal Account. *Stroke, 32*(11), 2719–2724. https://doi.org/10.1161/hs1101.098765

FISHER, M. (1951). OCCLUSION OF THE INTERNAL CAROTID ARTERY. *Archives of Neurology And Psychiatry*, *65*(3), 346. https://doi.org/10.1001/archneurpsyc.1951.02320030083009

FISHER, M., & ADAMS, R. D. (1951). Observations on brain embolism with special reference to the mechanism of hemorrhagic infarction. *Journal of Neuropathology and Experimental Neurology*, *10*(1), 92–94.

Kubicki, K., & Grzybowski, A. (2022). Pioneers in neurology: Charles Miller Fisher (1913–2012). *Journal of Neurology*, *269*(3), 1727–1729. https://doi.org/10.1007/s00415-021-10760-x

Noioso, C. M., Bevilacqua, L., Acerra, G. M., della Valle, P., Serio, M., Vinciguerra, C., Piscosquito, G., Toriello, A., Barone, P., & Iovino, A. (2023). Miller Fisher syndrome: an updated narrative review. *Frontiers in Neurology*, *14*. https://doi.org/10.3389/fneur.2023.1250774

Ojemann, R. G. (1984). Biography of C. Miller Fisher, M.D. *Neurosurgery*, *31*(Supplement 1), xiii–xviii. https://doi.org/10.1093/neurosurgery/31.CN_suppl_1.xiii

Chapter 9 – Forgotten

Cipriani, G., Dolciotti, C., Picchi, L., & Bonuccelli, U. (2011). Alzheimer and his disease: a brief history. *Neurological Sciences*, *32*(2), 275–279. https://doi.org/10.1007/s10072-010-0454-7

Goedert, M., & Ghetti, B. (2007). Alois Alzheimer: His Life and Times. *Brain Pathology*, *17*(1), 57–62. https://doi.org/10.1111/j.1750-3639.2007.00056.x

Hippius, H., & Neundörfer, G. (2003). The discovery of Alzheimer's disease. *Dialogues in Clinical Neuroscience*, *5*(1), 101–108. https://doi.org/10.31887/DCNS.2003.5.1/hhippius

Maurer, K., Volk, S., & Gerbaldo, H. (1997). Auguste D and Alzheimer's disease. *The Lancet*, *349*(9064), 1546–1549. https://doi.org/10.1016/S0140-6736(96)10203-8

Möller, H.-J., & Graeber, M. B. (1998). The case described by Alois Alzheimer in 1911. *European Archives of Psychiatry and Clinical Neuroscience*, *248*(3), 111–122. https://doi.org/10.1007/s004060050027

Page, S., & Fletcher, T. (2006). Auguste D: One hundred years on: 'The person 'not 'the case.' *Dementia*, *5*(4), 571–583. https://doi.org/10.1177/1471301206069939

Scheltens, P., de Strooper, B., Kivipelto, M., Holstege, H., Chételat, G., Teunissen, C. E., Cummings, J., & van der Flier, W. M. (2021). Alzheimer's disease. *Lancet (London, England)*, *397*(10284), 1577–1590. https://doi.org/10.1016/S0140-6736(20)32205-4

Tagarelli, A., Piro, A., Tagarelli, G., Lagonia, P., & Quattrone, A. (2006). Alois Alzheimer: a hundred years after the discovery of the eponymous disorder. *International Journal of Biomedical Science : IJBS*, *2*(2), 196–204.

Image:
File:Auguste D aus Marktbreit.jpg. (2023, October 2). *Wikimedia Commons*. Retrieved January 6, 2025 from https://commons.wikimedia.org/w/index.php?title=File:Auguste_D_aus_Marktbreit.jpg&oldid=807424419.

Chapter 10 – The Showman

Beeson, B. B. (1928). Jean Martin Charcot: A Summary of His Life and Works. *Annals of Medical History*, *10*(2), 126–132.

Camargo, C. H. F., Coutinho, L., Neto, Y. C., Engelhardt, E., Filho, P. M., Walusinski, O., & Teive, H. A. G. (2023). Jean-Martin Charcot: the polymath. *Arquivos de Neuro-Psiquiatria*, *81*(12), 1098–1111. https://doi.org/10.1055/s-0043-1775984

Cassady, M. (2019). Hysteria to Functional Neurologic Disorders: A Historical Perspective. *American Journal of Psychiatry Residents' Journal*, *15*(1), 15–15. https://doi.org/10.1176/appi.ajp-rj.2019.150111

Clanet, M. (2008). Jean-Martin Charcot. 1825 to 1893. *International MS Journal, 15*(2), 59–61.

Giménez-Roldán, S. (2016). Clinical history of Blanche Wittman and current knowledge of psychogenic non-epileptic seizures. *Neurosciences and History, 4*(4), 122–129.

Kumar, D. R., Aslinia, F., Yale, S. H., & Mazza, J. J. (2011). Jean-Martin Charcot: The Father of Neurology. *Clinical Medicine & Research, 9*(1), 46–49. https://doi.org/10.3121/cmr.2009.883

Teive, H. A. G., Arruda, W. O., & Werneck, L. C. (2005). Rosalie: the brazilian female monkey of Charcot. *Arquivos de Neuro-Psiquiatria, 63*(3a), 707–708. https://doi.org/10.1590/S0004-282X2005000400031

Thorburn, A. L. (1967). Jean Martin Charcot, 1825-1893. An appreciation. *Sexually Transmitted Infections, 43*(2), 77–80. https://doi.org/10.1136/sti.43.2.77

Walusinski, O. (2014). *The Girls of La Salpêtrière* (pp. 65–77). https://doi.org/10.1159/000359993

White, M. B. (1997). Jean-Martin Charcot's Contributions to the Interface Between Neurology and Psychiatry. *Canadian Journal of Neurological Sciences / Journal Canadien Des Sciences Neurologiques, 24*(3), 254–260. https://doi.org/10.1017/S0317167100021909

Image:

File:Une leçon clinique à la Salpêtrière.jpg. (2024, March 20). *Wikimedia Commons.* Retrieved January 6, 2025 from https://commons.wikimedia.org/w/index.php?title=File:Une_le%C3%A7on_clinique_%C3%A0_la_Salp%C3%AAtri%C3%A8re.jpg&oldid=862144171.

Chapter 11 – Storybook Syndromes

Alice in Wonderland Syndrome

Blom, J. D. (2016). Alice in Wonderland syndrome. *Neurology Clinical Practice, 6*(3), 259–270. https://doi.org/10.1212/CPJ.0000000000000251

Carroll, L. (1865). *Alice's Adventures in Wonderland.* MacMillan and Co.

Kitchener, N. (2004). Alice in Wonderland syndrome. *Int J Child Neuropsychiatry, 1*(1), 107–112.

Mastria, G., Mancini, V., Viganò, A., & di Piero, V. (2016). Alice in Wonderland Syndrome: A Clinical and Pathophysiological Review. *BioMed Research International, 2016,* 1–10. https://doi.org/10.1155/2016/8243145

O'Toole, P., & Modestino, E. J. (2017). Alice in Wonderland Syndrome: A real life version of Lewis Carroll's novel. *Brain and Development, 39*(6), 470–474. https://doi.org/10.1016/j.braindev.2017.01.004

Munchausen Syndrome

Asher, R. (1951). MUNCHAUSEN'S SYNDROME. *The Lancet, 257*(6650), 339–341. https://doi.org/10.1016/S0140-6736(51)92313-6

Hariharasubramony, A., Chankramath, S., & Srinivasa, S. (2012). Munchausen Syndrome as Dermatitis Simulata. *Indian Journal of Psychological Medicine, 34*(1), 94–96. https://doi.org/10.4103/0253-7176.96171

Meadow, R. (1989). ABC of child abuse. Munchausen syndrome by proxy. *British Medical Journal, 299*(6693), 248–250. https://doi.org/10.1136/bmj.299.6693.248

Sousa Filho, D. de, Kanomata, E. Y., Feldman, R. J., & Maluf Neto, A. (2017). Munchausen syndrome and Munchausen syndrome by proxy: a narrative review. *Einstein (São Paulo), 15*(4), 516–521. https://doi.org/10.1590/s1679-45082017md3746

Othello Syndrome

Cipriani, G., Vedovello, M., Nuti, A., & di Fiorino, A. (2012). Dangerous passion: Othello syndrome and dementia. *Psychiatry and Clinical Neurosciences, 66*(6), 467–473. https://doi.org/10.1111/j.1440-1819.2012.02386.x

Park, J. H., Sarwar, S., Hassett, L. C., Staab, J. P., & Fipps, D. C. (2024). Clinical Characterization, Course, and Treatment of Othello Syndrome: A Case Series and Systematic Review of the Literature. *Journal of the Academy of Consultation-Liaison Psychiatry, 65*(1), 89–105. https://doi.org/10.1016/j.jaclp.2023.09.006

Shakespeare, W. (1868). *Othello the Moor of Venice* (Vol. 31). B. Tauchnitz.

Pickwickian Syndrome

Burwell, C. S., Robin, E. D., Whaley, R. D., & Bickelmann, A. G. (1994). Extreme Obesity Associated with Alveolar Hypoventilation—A Pickwickian Syndrome*. *Obesity Research, 2*(4), 390–397. https://doi.org/10.1002/j.1550-8528.1994.tb00084.x

Dickens, C. (1838). *The posthumous papers of the Pickwick Club.* Carey, Lea and Blanchard.

Ghimire, P., Sankari, A., & Kaul, P. (2024). *Pickwickian Syndrome.* StatPearls [Internet]. https://www.ncbi.nlm.nih.gov/books/NBK542216/

Littleton, S. W., & Mokhlesi, B. (2009). The Pickwickian Syndrome—Obesity Hypoventilation Syndrome. *Clinics in Chest Medicine, 30*(3), 467–478. https://doi.org/10.1016/j.ccm.2009.05.004

Rapunzel Syndrome

Altonbary, A. Y., & Bahgat, M. H. (2015). Rapunzel syndrome. *Journal of Translational Internal Medicine, 3*(2), 79–81. https://doi.org/10.1515/jtim-2015-0008

Naik, S., Gupta, V., Naik, S., Rangole, A., Chaudhary, A. K., Jain, P., & Sharma, A. K. (2007). Rapunzel Syndrome Reviewed and Redefined. *Digestive Surgery, 24*(3), 157–161. https://doi.org/10.1159/000102098

Vaughan, E. D., Sawyers, J. L., & Scott, H. W. (1968). The Rapunzel syndrome. An unusual complication of intestinal bezoar. *Surgery, 63*(2), 339–343.

Chapter 12 – The Art and Anatomy of War

Darwin, C. (1872). *The Expression of the Emotions in Man and Animals.* P. Ekman, Ed.

Gijn, J. (2011). Charles Bell (1774–1842). *Journal of Neurology, 258*(6), 1189–1190. https://doi.org/10.1007/s00415-011-5912-5

Gordon-Taylor, G., & Wright Walls, E. (1958). *Sir Charles Bell: His Life and Times.* E. & S. Livingstone.

Grzybowski, A., & Kaufman, M. H. (2007). Sir Charles Bell (1774–1842): contributions to neuro-ophthalmology. *Acta Ophthalmologica Scandinavica, 85*(8), 897–901. https://doi.org/10.1111/j.1600-0420.2007.00972.x

Kazi, R. A., & Rhys-Evans, P. (2004). Sir Charles Bell: The artist who went to the roots! *Journal of Postgraduate Medicine, 50*(2), 158–159.

Tiemstra, J. D., & Khatkhate, N. (2007). Bell's palsy: diagnosis and management. *American Family Physician, 76*(7), 997–1002.

Images:

File:Watercolour, wounded soldier at Waterloo Wellcome L0022544.jpg. (2020, February 19). *Wikimedia Commons*. Retrieved January 6, 2025 from https://commons.wikimedia.org/w/index.php?title=File:Watercolour,_wounded_soldier_at_Waterloo_Wellcome_L0022544.jpg&oldid=395984321.

File:Watercolour of wounded soldier, Waterloo, 1815 Wellcome L0022539.jpg. (2020, February 19). *Wikimedia Commons*. Retrieved January 6, 2025 from https://commons.wikimedia.org/w/index.php?title=File:Watercolour_of_wounded_soldier,_Waterloo,_1815_Wellcome_L0022539.jpg&oldid=395984322.

File:Watercolour, wounded soldier at Waterloo Wellcome L0022548.jpg. (2020, February 19). *Wikimedia Commons*. Retrieved January 6, 2025 from https://commons.wikimedia.org/w/index.php?title=File:Watercolour,_wounded_soldier_at_Waterloo_Wellcome_L0022548.jpg&oldid=395984318.

Chapter 13 – The Dancing Curse

Bhattacharyya, K. (2016). The story of George Huntington and his disease. *Annals of Indian Academy of Neurology, 19*(1), 25. https://doi.org/10.4103/0972-2327.175425

Cubo, E. (2016). Huntington Disease: A Journey through History. *Neurosci Hist, 4*(4), 160–163.

Durbach, N., & Hayden, M. R. (1993). George Huntington: the man behind the eponym. *Journal of Medical Genetics, 30*(5), 406–409. https://doi.org/10.1136/jmg.30.5.406

McColgan, P., & Tabrizi, S. J. (2018). Huntington's disease: a clinical review. *European Journal of Neurology, 25*(1), 24–34. https://doi.org/10.1111/ene.13413

Owecki, M. K., & Magowska, A. (2019). George Huntington (1850–1916). *Journal of Neurology, 266*(3), 793–795. https://doi.org/10.1007/s00415-018-8860-5

Wexler, A., Wild, E. J., & Tabrizi, S. J. (2016). George Huntington: a legacy of inquiry, empathy and hope. *Brain, 139*(8), 2326–2333. https://doi.org/10.1093/brain/aww165

Chapter 14 – In the Family

Boland, C. R. (2019). Henry T. Lynch, MD (January 4, 1928–June 2, 2019). *Gastroenterology, 157*(4), 905–906. https://doi.org/10.1053/j.gastro.2019.08.029

Cantor, D. (2006). The Frustrations of Families: Henry Lynch, Heredity, and Cancer Control, 1962-1975. *Medical History, 50*(3), 279–302.

Kolata, G. (2019, June 13). Dr. Henry Lynch, 91, Dies; Found Hereditary Link in Cancer. *The New York Times*.

Langer, E. (2019, June 4). Henry Lynch, celebrated as father of cancer genetics, dies at 91. *The Washington Post*.

Marcus, A. (2019). Henry T Lynch. *The Lancet, 394*(10192), 22. https://doi.org/10.1016/S0140-6736(19)31502-8

Sehgal, R., Sheahan, K., O'Connell, P., Hanly, A., Martin, S., & Winter, D. (2014). Lynch Syndrome: An Updated Review. *Genes, 5*(3), 497–507. https://doi.org/10.3390/genes5030497

The Oncology Nurse. (2013, March). *Lynch Syndrome: An Interview With the Father of Hereditary Cancer Detection and Prevention, Henry T. Lynch, MD*. The Oncology Nurse. https://theoncologynurse.com/interview-with-the-innovators/15640-lynch-syndrome-an-interview-with-the-father-of-hereditary-cancer-detection-and-prevention-henry-t-lynch-md

Warren, P. (2019, June 21). Dr Henry Lynch obituary. *The Guardian*.

Chapter 15 – One of Us

Antonarakis, S. E., Skotko, B. G., Rafii, M. S., Strydom, A., Pape, S. E., Bianchi, D. W., Sherman, S. L., & Reeves, R. H. (2020). Down syndrome. *Nature Reviews Disease Primers*, *6*(1), 9. https://doi.org/10.1038/s41572-019-0143-7

Down, J. L. H. (1866). Observations on an Ethnic Classification of Idiots. *London Hospital Reports*, *3*(1866), 259–262.

Dunn, P. M. (1991). Dr Langdon Down (1828-1896) and "mongolism". *Archives of Disease in Childhood*, *66*(7 Spec No), 827–828. https://doi.org/10.1136/adc.66.7_Spec_No.827

Kutzsche, S. (2018). John Langdon Down (1828–1896) – a pioneer in caring for mentally disabled patients. *Acta Paediatrica*, *107*(11), 1851–1854. https://doi.org/10.1111/apa.14505

van Robays, J. (2016). John Langdon Down (1828 - 1896). *Facts, Views & Vision in ObGyn*, *8*(2), 131–136.

Ward, O. C. (1999). John Langdon Down: The Man and the Message. *Down Syndrome Research and Practice*, *6*(1), 19–24. https://doi.org/10.3104/perspectives.94

Image:

File:Photograph of a patient at Earlswood Asylum by John Langdon Down.jpg. (2015, June 26). Langdon Down Museum of Learning Disability. Retrieved January 6, 2025, from https://langdondownmuseum.org.uk/2015/06/26/u3a-project-2014-long-stay-institutions-for-people-with-learning-disabilities-earlswood-asylum/

Chapter 16 – The Battlefield of the Brain

Bliss, M. (2005). *Harvey Cushing: A Life in Surgery*. Oxford University Press.

Cannon, W. B. (1941). Harvey (Williams) Cushing 1869-1939. *Obituary Notices of Fellows of the Royal Society*, *3*(9), 277–290. https://doi.org/10.1098/rsbm.1941.0003

Doyle, N. M., Doyle, J. F., & Walter, E. J. (2017). The life and work of Harvey Cushing 1869–1939: A pioneer of neurosurgery. *Journal of the Intensive Care Society*, *18*(2), 157–158. https://doi.org/10.1177/1751143716673076

Ellis, H. (2012). Harvey Cushing: Cushing's Disease. *Journal of Perioperative Practice*, *22*(9), 298–299. https://doi.org/10.1177/175045891202200906

Fulton, J. F. (1954). Harvey Cushing As We Knew Him. *Bulletin of the New York Academy of Medicine*, *30*(11), 886–915.

Kennedy, J. C. (1941). Harvey Cushing (1869-1939). *University of Western Ontario Medical Journal*, *11*(4), 125–138.

Lacroix, A., Feelders, R. A., Stratakis, C. A., & Nieman, L. K. (2015). Cushing's syndrome. *The Lancet*, *386*(9996), 913–927. https://doi.org/10.1016/S0140-6736(14)61375-1

Chapter 17 – A Dubious Honour

Carrión's disease

Cueto, M. (1996). Tropical medicine and bacteriology in Boston and Peru: Studies of Carrión's disease in the early twentieth century. *Medical History*, *40*(3), 344–364. https://doi.org/10.1017/S0025727300061330

Stürup, A. R. S. (2023). Daniel Alcides Carrión García. *Cazadores de Microbios En Venezuela y El Mundo*.

Lou Gehrig's disease

ALS Association. (n.d.). *Lou Gehrig and the History of ALS*. Retrieved January 3, 2025, from https://www.als.org/understanding-als/lou-gehrig

Backer, R. (2020). Lou Gehrig, Movie Star. *The Baseball Research Journal, 49*(2), 14–19.

Brennan, F. (2012). The 70th Anniversary of the Death of Lou Gehrig. *American Journal of Hospice and Palliative Medicine®, 29*(7), 512–514. https://doi.org/10.1177/1049909111434635

Kasarskis, E. J. (2006). A New Perspective on the Life of Lou Gehrig. *Neurology Today, 6*(11), 21.

Lewis, M., & Gordon, P. H. (2007). Lou Gehrig, Rawhide, and 1938. *Neurology, 68*(8), 615–618. https://doi.org/10.1212/01.wnl.0000254623.04219.aa

National Baseball Hall of Fame. (n.d.). *Luckiest Man*. Retrieved January 3, 2025, from https://baseballhall.org/discover-more/stories/baseball-history/lou-gehrig-luckiest-man

Robinson, R. (2005, July 3). Gehrig Remains a Presence in His Former Neighborhood. *The New York Times*.

Christmas disease

Biggs, R., Douglas, A. S., Macfarlane, R. G., Dacie, J. v., Pitney, W. R., Merskey, C., & O'Brien, J. R. (1952). Christmas Disease. *BMJ, 2*(4799), 1378–1382. https://doi.org/10.1136/bmj.2.4799.1378

Biggs, R., & Macfarlane, R. G. (1962). Christmas Disease. *Postgraduate Medical Journal, 38*(435), 3–12. https://doi.org/10.1136/pgmj.38.435.3

Giangrande, P. L. F. (2003). Six Characters in Search of An Author: The History of the Nomenclature of Coagulation Factors. *British Journal of Haematology, 121*(5), 703–712. https://doi.org/10.1046/j.1365-2141.2003.04333.x

Kopplin, P. (2020, January 22). *From eponym to advocate: The story of Stephen Christmas*. Hektoen International A Journal of Medical Humanities. https://hekint.org/2020/01/22/from-eponym-to-advocate-the-story-of-stephen-christmas/

Chapter 18 – The Silent Dissenter

Carrilho, P. E. M., & Nitrini, R. (2021). The controversial Third Reich history of Hans Creutzfeldt: was he a supporter or just another adept of the "hand washing policy"? *Arquivos de Neuro-Psiquiatria, 79*(1), 84–87. https://doi.org/10.1590/0004-282 x-anp-2020-0274

Duckett, S., & Stern, J. (1999). Origins of the Creutzfeldt and Jakob Concept. *Journal of the History of the Neurosciences, 8*(1), 21–34. https://doi.org/10.1076/jhin.8.1.21.1771

Illert, M., & Schmidt, M. (2020). Hans Gerhard Creutzfeldt (1885–1964) in the Third Reich. *Neurology, 95*(2), 72–76. https://doi.org/10.1212/WNL.0000000000009785

Iwasaki, Y. (2017). Creutzfeldt-Jakob disease. *Neuropathology, 37*(2), 174–188. https://doi.org/10.1111/neup.12355

Sammet, K. (2008). Alfons Jakob (1884–1931). *Journal of Neurology, 255*(11), 1852–1853. https://doi.org/10.1007/s00415-008-0918-3

Triarhou, L. C. (2009). Alfons Maria Jakob (1884–1931), Neuropathologist par Excellence. *European Neurology, 61*(1), 52–58. https://doi.org/10.1159/000175123

United States Holocaust Memorial Museum. (2020, October 7). *Euthanasia Program and Aktion T4*. Holocaust Encyclopedia. https://encyclopedia.ushmm.org/content/en/article/euthanasia-program

Wolf, J. H., & Foley, P. (2005). Hans Gerhard Creutzfeldt (1885–1964): a life in neuropathology. *Journal of Neural Transmission, 112*(8), I–XCVII. https://doi.org/10.1007/s00702-005-0288-2

Chapter 19 – Myths, Madness, Ticks and Tics
Greek Mythology & Medicine

Freud, S. (1924). The Passing of the Oedipus Complex. *International Review of Psycho-Analysis*, 5, 419–424.

Karenberg, A. (2013). The world of gods and the body of man: mythological origins of modern anatomical terms. *Anatomy (International Journal of Experimental and Clinical Anatomy)*, 6–7, 7–22. https://doi.org/10.2399/ana.11.142

Kucharz, E. J. (2017). Medical Eponyms of Mythological Origin. *Acta Neophilologica*, 2(XIX), 29–42.

Mad Cow Disease

Bovine Spongiform Encephalopathy (BSE). (2024, May 10). Centers for Disease Control and Prevention. https://www.cdc.gov/mad-cow/php/animal-health/index.html

Casalone, C., & Hope, J. (2018). *Atypical and classic bovine spongiform encephalopathy* (pp. 121–134). https://doi.org/10.1016/B978-0-444-63945-5.00007-6

Fisher, J. R. (1998). Cattle Plagues Past and Present: The Mystery of Mad Cow Disease. *Journal of Contemporary History*, 33(2), 215–228. https://doi.org/10.1177/002200949803300202

Nathanson, N., Wilesmith, J., & Griot, C. (1997). Bovine Spongiform Encephalopathy (BSE): Causes and Consequences of a Common Source Epidemic. *American Journal of Epidemiology*, 145(11), 959–969. https://doi.org/10.1093/oxfordjournals.aje.a009064

Pain, S. (1987, November 5). Brain disease drives cows wild. *New Scientist*.

Lyme Disease

Elbaum-Garfinkle, S. (2011). Close to home: a History of Yale and Lyme Disease. *The Yale Journal of Biology and Medicine*, 84(2), 103–108.

Stanek, G., Wormser, G. P., Gray, J., & Strle, F. (2012). Lyme borreliosis. *The Lancet*, 379(9814), 461–473. https://doi.org/10.1016/S0140-6736(11)60103-7

Steere, A. C., Strle, F., Wormser, G. P., Hu, L. T., Branda, J. A., Hovius, J. W. R., Li, X., & Mead, P. S. (2016). Lyme borreliosis. *Nature Reviews Disease Primers*, 2(1), 16090. https://doi.org/10.1038/nrdp.2016.90

TikTok Tics

Hull, M., & Parnes, M. (2021). Tics and TikTok: Functional Tics Spread Through Social Media. *Movement Disorders Clinical Practice*, 8(8), 1248–1252. https://doi.org/10.1002/mdc3.13267

Olvera, C., Stebbins, G. T., Goetz, C. G., & Kompoliti, K. (2021). TikTok Tics: A Pandemic Within a Pandemic. *Movement Disorders Clinical Practice*, 8(8), 1200–1205. https://doi.org/10.1002/mdc3.13316

Shmerling, R. H. (2022, January 18). Tics and TikTok: Can social media trigger illness? *Harvard Health Publishing*. https://www.health.harvard.edu/blog/tics-and-tiktok-can-social-media-trigger-illness-202201182670

Printed in Dunstable, United Kingdom

72314195R00211